THE HAIDER PHENOMENON

BY
MELANIE A. SULLY

EAST EUROPEAN MONOGRAPHS
DISTRIBUTED BY COLUMBIA UNIVERSITY PRESS, NEW YORK

1997

EAST EUROPEAN MONOGRAPHS, NO. CDLXXXIV

Copyright © 1997 by Melanie A. Sully
ISBN 0-88033-381-2
Library of Congresss Catalog Card Number 97-61667

Printed in the United States of America

TABLE OF CONTENTS

PREFACE		ix
JÖRG HAIDER: A BRIEF BIOGRAPHY		xi
1.	THE GLADIATOR	1
2.	ORGANISATION '95	13
	The Citizens' Movement 1998	15
	Contract with Austria	17
3.	THE "THIRD REPUBLIC"	21
	"The Country Must Change"	23
4.	"IDEOLOGY"	35
	Trans-Atlantic Trip	40
5.	PROGRAMMATIC DEVELOPMENT	44
	"Austria über alles"	53
	"Onward Christian Soldiers"	55
	A Free Republic	56
6.	HAIDER'S CREDO	57
	Vienna Declaration	60
	The Austria Declaration	62
	The German Question	63
	"Adolf Haider?"	66
	"Penal Camps"	67
	The Centre Fringe	69
	"Penal Camps" Speech	73
7.	"AUSTRIA FIRST"	77
	12 Points of the Popular Initiative 1993	87
8.	EUROSCEPTICISM	91
	Security	98
	Election '94	101

9. ELECTION '95	107
Background to the '95 Election	108
Budget Troubles	109
The Campaign	110
The Results	117
20 Pledges for the "Contract with Austria"	122
10. TO KRUMPENDORF AND BACK	128
The Krumpendorf Speech	138
11. F – WHAT NOW?	143
Third President	149
All Change	156
Forty Years of the "FPÖ"	158
The Tide Turns	161
The Spectre of Terrorism	163
12. EURO ELECTIONS '96 AND AFTERMATH	169
"Vienna must not become Chicago"!	170
Black and White	172
Maastricht to Waterloo	174
Feldkirch '96	177
"Red Bank – Black Bank"	180
Vranitzky "adieu"	186
13. THE ENTERTAINER	199
INTERVIEWS	
Kurt Waldheim	206
Otto Habsburg	208
Jörg Haider	211
Simon Wiesenthal	221
APPENDICES	
1. Background Information	224
2. Brief Profiles of Other Main Parties	228

FIGURES

Figure 1	Party Composition of Post-war Governments	6
Figure 2	FPÖ Share of the Vote in Elections to the National Council	9
Figure 3	Citizens' Movement '98	19
Figure 4	Contract with Austria	20
Figure 5	Combined Share of Vote, ÖVP and SPÖ, 1945-1995	119
Figure 6	Constitutional Bow	166
Figure 7	Electoral Behaviour in European Parliament Election 1996 Compared to National Council Election 1995 (Men)	191
Figure 8	Electoral Behaviour in European Parliament Election 1996 Compared to National Council Election 1995 (Women)	192
Figure 9	Electoral Behaviour in European Parliament Election 1996 Compared to National Council Election 1995 (Young People under 30)	193
Figure 10	Electoral Behaviour in European Parliament Election 1996 Compared to National Council Election 1995 (Blue-collared Workers)	194
Figure 11	Electoral Behaviour in European Parliament Election 1996 Compared to National Council Election 1995 (Salaried Employees/Civil Servants)	195
Figure 12	Seats in Vienna Municipality, 1996	196
Figure 13	Municipal Elections in Vienna 1945-96	197

Figure A1	Stages of Legislation	225
Figure A2	Electoral Profile of the Parties 1945-1995	226
Figure A3	Election Results for the Main Parties (National Council Elections; European Parliament Election, 96)	227
Figure A4	Public Perception of Position of Parties	235
Figure A5	SPÖ Membership 1945-1995	236
Figure A6	SPÖ Membership 1985-1995	237

TABLES

Table 1	Election Results to the National Council, 1945-1995	7
Table 2	National Council Elections, 1949-1995, Results for the WdU (Independents) and FPÖ	8
Table 3	Composition of the Federal Council, 1945-96	10
Table 4	FPÖ Provincial Election Results, 1981-1996	11
Table 5	Popular Initiatives in Austria, 1945-1997	86
Table 6	Results of the EU Referendum, 1994	103
Table 7	EU Referendum Voting Behaviour (%)	104
Table 8	Socio-demographic Breakdown of Votes for the FPÖ, 1986-1995	120
Table 9	Changes in Electoral Behaviour in National Council Elections, 1986-1995	121

Table 10	Final Result of the European Parliament Election, 13 October 1996	189
Table 11	The 1996 European Parliament Elections: A Socio-Demographic Analysis	190
Table 12	FPÖ Membership 1992-1996	198
Table A1	Chancellors of the Second Republic	224

PHOTOGRAPHS
(Following Chapter 7)

Dust jacket: Confident of winning (Photo: Eggenberger)

1. Relaxing (Photo: Holzner)

2. 1974 Press conference, as leader of Free Youth of Austria (Photo: Fritz Kern)

3. Election as leader of FPÖ, Innsbruck 1986. (Photo: Holzner)

4. With Franz Vranitzky (Photo: Holzner)

5. Voting with his wife (Photo: Holzner)

6. With an estranged Heide Schmidt (Photo: Holzner)

7. With President Kurt Waldheim (Photo: Holzner)

8. With Niki Lauda, racing driver and airline owner (Photo: Fritz)

9. The FPÖ moves towards the Church (Photo: Fritz)

10. Thousands demonstrate in Klagenfurt in protest against Haider's dismissal as provincial governor in 1991 (Photo: Fritz)

11. Addressing the Ulrichsberg Festival (Photo: Fritz)

12. Rock climbing (Photo: Fritz)

13. Victorious at sea (Photo: Eggenberger)

14. Ewald Stadler, Haider's number 2 in parliament; nicknamed "the Doberman" (Photo: Holzner)

15. Looking into the future (Photo: Fritz)

Preface

This book should be of interest to those seeking more information about the most controversial politician in Austria today - Jörg Haider, leader of one of the most successful right wing parties in Europe. It deals with recent political developments and includes information on the programmatic position of the modern Freedom Party of Austria (FPÖ).

Haider's statements on Hitler's Third Reich and the Austrian nation have provoked criticism both at home and abroad. Some believe if Haider came to power he would do more to damage Austria's international image than former president Kurt Waldheim.

At the last election in 1995, Haider's party won almost 22 percent of the vote and in the elections to the European parliament in October 1996 it surged to gain 27.5 percent putting it on almost equal footing with the two governing parties.

Haider took over as party leader in September 1986 at a stormy party conference in Innsbruck. Then the party could reckon with around 250,000 votes; today it can boast over a million. Alarmingly for the Social Democrats, Haider has made great inroads in working-class districts where his rhetoric on immigration and unemployment has come home. Now in Austria there are three medium-sized parties, including the Christian Democratic People's Party, which all stand a good chance of emerging as the largest force at the next election scheduled to be held at the latest by the end of 1999.

Haider has declared he will be federal chancellor by the end of the century and has developed a "Contract with Austria," based on the Newt Gingrich model, which he pledges to put into practice. He restyled his party into an electoral movement with the purpose of getting him into the Vienna Ballhausplatz as chancellor of Austria.

This book documents Haider's recent policy statements and his changing plans for a "Third Republic." It looks at topical issues such as immigration, the European Union, constitutional reform and attitudes to the Nazi past and the Holocaust. It provides information on the electoral success of the party and examines some of the most recent controversies surrounding Haider and his Freedomites.

The Haider party is the pariah of modern politics in Austria. It is often reviled and shunned as "semi-fascist" and scorned for its alleged contacts with the neo-Nazi fringe.

Much of the time and energy of the other parties has been directed at "stopping Haider," a strategy which seems to have miserably failed. This book considers the very real changes which are needed in Austria today and reflects on the contribution of Jörg Haider and the FPÖ to this debate.

Finally it includes interviews with Kurt Waldheim, Otto Habsburg, Simon Wiesenthal and Haider himself, as well as photographs of Haider's life and political career.

••••

Many people have been kind enough to help me with this book including the staff of Haider's office, the *Bündnisbüro* and the *Freiheitliche Akademie*.

I am grateful to Manfried Welan, Lothar Höbelt, Christine Young and Walter Howadt for their perceptive comments and suggestions.

Finally, as always, I am indebted to dear Rog for his support and encouragement over many years.

Vienna
Easter 1997

JÖRG HAIDER: A BRIEF BIOGRAPHY

26 January 1950	Born in Bad Goisern, Upper Austria; father – shoemaker; former member of the Nazi party; served in the German army during the war; mother former member of Hitler's League of German Maidens
1966	Winner of debating contest with a talk entitled "Are we Austrians German?" Leader of the FPÖ youth movement in Upper Austria
1968	Military service in the federal army
1969	University studies (law) in Vienna, doctorate 1973
1970	Member of the Federal Army Committee on reform
1973-76	Assistant professor at the Institute for State and Administrative Law at the University of Vienna
1970-74	Leader of the Ring of Free Youth, Austria
since 1974	Member of the FPÖ party executive
1976-83	Party secretary of the FPÖ for the province of Carinthia
since 1983	Leader of the FPÖ in Carinthia
1979-83	Member of parliament and spokesman for social affairs (regarded as a "liberal")
1982-3	Chief editor of the *Kärntner Nachrichten*
1983-86	FPÖ member of the provincial diet in Carinthia with responsibility for tourism and road construction
13 September 1986	Elected as leader of the FPÖ at the party conference in Innsbruck
1986-89	Member of parliament
1989	Triumph at the provincial elections in Carinthia and elected as provincial governor with the votes of the FPÖ and ÖVP.
21 June 1991	Loss of governorship by a vote in the Carinthian diet
1992	Member of parliament; floor leader of the FPÖ parliamentary party.

Jörg Haider is married and has two daughters; his elder sister is also an FPÖ politician. His interests include mountaineering, rock climbing, bungee jumping, tennis, skiing, marathon running (New York – 3 hours 52 minutes; Vienna – 3 hours 31 minutes). He is the author of *Die Freiheit die ich Meine* (Ullstein: Frankfurt/Main, 1993) and *Friede durch Sicherheit* (Freiheitliche Akademie: Vienna, 1996).

For further biographical details see: *Jörg im Bild* (FP-1998, Hollabrunn, 1996).

1

THE GLADIATOR

With Haider a modern political culture has emerged in Austrian politics, more strident and less consensus oriented than the founding fathers of the Second Republic. Born in 1950, Jörg Haider was the product of a different generation, more inclined to challenge old orthodoxy and taboos. He was to become a polariser, not a reconciler so typical of the Second Republic, but a politician whose diction bordered on and occasionally went beyond what was considered politically decent. This was a novel phenomenon for Austria and the old school seemed unsure how to handle it. Haider's opponents reacted with increasing nervousness as his popularity grew. Most seemed helpless in dealing with this new brand of right radical populism with its sound-bite rhetoric which went directly to the heart of people's needs and worries. Attempts to label Haider as a destructive extremist were crude and counter-productive. Deep down even his adversaries knew that Haider was only voicing what many felt but could not openly express.

An excessive concentration on the personality of Haider transformed him into a kind of political "Terminator," larger than life. Even serious writers analyse changes in his hair style, the kinds of shoes he wears and even irregularities in his teeth. Many indulged in this while lamenting the "Americanisation" of politics with its emphasis on glossy packaging of the individual instead of serious intellectual discussion. Despite this, a debate has emerged over the last years in Austria which has called into question many institutions hitherto regarded as "untouchable." Privileges for politicians, party patronage and "jobs for the boys" are more easily exposed, and less tolerated by an increasingly critical public. Haider has not been the only factor contributing to this sea change, but he had the good fortune to be in the right place at the right time and to be blessed with the consummate art of communication. Over the last ten years his FPÖ has been transformed from a party fearing extinction to one which aspires to power as the leading force in the country. His opponents no longer ridicule this ambition but fear it could become true and herald the end of the Second Republic.

The present Freedom Party of Austria under Jörg Haider is the heir to the pan-German nationalist and liberal camp which had its roots in the nineteenth century. It was traditionally anti-Slav and anti-clericist in outlook and shared little in common with the Anglo-American idea of liberalism. Some of its protagonists, such as Georg von Schönerer, were radically anti-Semitic but there were many conflicting factions in this camp.[1] This "third force" in Austrian politics has enjoyed a chequered history in the post 1945 period. Initially banned by the Allies from political activity, the pan-German nationalists gravitated around an electoral group which was allowed to contest the 1949 election. The Socialist SPÖ welcomed the appearance of the newcomers in the belief that this would break any "bourgeois bloc." The Electoral League of Independents (WdU) gained 11.6 percent of the total votes which gave them 16 seats in parliament inflicting damage on the Socialists as well as the Right. This party included many diverse elements and some of its founding members had spoken out against Nazism and had been imprisoned during the Third Reich. Ex-Nazis were courted not only by the Independents but also by the main parties after the war.

The Second Republic, founded in 1945, was dominated by the "Reds" (the SPÖ) and the "conservative" People's Party (ÖVP) known as the "Blacks" (see Figure 1). Together they developed a power network which encompassed many facets of economic, social, cultural and political life. Coalition government was reinforced through social partnership and *Proporz* with its politicised appointments in nationalised industries, banks, hospitals and schools.[2] The Great Coalition provided stability in contrast to the centrifugal forces of the First Republic which had crashed in a fratricidal civil war between the predecessors of the Blacks and Reds. It also generated prosperity and confidence in a new democratic republic which eventually won sovereignty and independence in 1955 after ten years of allied occupation.

In 1955/56 the Freedom Party of Austria (FPÖ) was formed as a successor to the "Independents" which had been weakened by internal disputes. A leading figure was Anton Reinthaller, a former prominent Nazi who had served in the 1938 *Anschluß* cabinet of Seyss-Inquart. The new party's electoral performance was moderate and it could not penetrate the hegemonial position of the big two (see Table 1). In the 1960s disillusionment grew with the functioning of the Great Coalition and the FPÖ became attractive as a potential partner. In 1963 the SPÖ joined forces with the FPÖ against the ÖVP in parliament in a move to prevent Otto Habsburg returning to Austria.

An electoral reform designed to help small parties was secured from Bruno Kreisky's socialist minority government at the beginning of the 1970s. The ÖVP had excluded the possibility of a coalition with the FPÖ on the eve of the 1970 election which paved the way for this collusion. In return the FPÖ gave qualified support for a short period from the opposition benches to Kreisky who was to be chancellor for the next thirteen years. The door had been opened to the FPÖ by Kreisky who had given the party a degree of respectability.

Throughout its history the FPÖ was haunted by the shadow of Nazism. The two streams in the party, pan-German nationalist and liberal, co-existed in an uneasy symbiotic relationship. The furor in the 1970s surrounding the leader Friedrich Peter, a former SS officer but known for his "liberal" outlook, highlighted the split personality of the party. Kreisky's support of Peter further underlined the complexity of Austria's brown past and the ghosts of the Third Reich.

At the end of the Kreisky era in 1983, the FPÖ entered government with the Socialists under chancellor Fred Sinowatz. This lasted just three years and proved a traumatic experience for the party then under the leadership of the lawyer, Norbert Steger, who also served as vice chancellor in the coalition. Steger steered the party to the left in an effort to give it a respectable "liberal" image but looked too often to be a socialist lapdog. The party's support dwindled alarmingly in the country and by autumn 1986 (just before Haider took over) it faced the prospect of political extinction at the polls (see Table 2 and Figure 2). The Carinthian party, where Jörg Haider had his power base, went its own way and defied the national trend. In provincial elections there in 1984, the FPÖ won 16 percent of the vote, a surprisingly good result. Frustration with the leadership in Vienna increased and calls for more radical policies came not only from Carinthia but also from Upper Austria and the city of Graz.

A series of scandals plagued the SPÖ involving some of its leading personalities. This brought politics and politicians into disrepute and was one factor behind increased electoral mobility, abstentions, and protest votes.[3]

Intra-party disputes came to a head at a stormy session during the Innsbruck conference of the FPÖ 13 September 1986 when Steger was toppled by his young challenger for the leadership, 36 year-old Jörg Haider, who won 58 percent of the delegates' votes. This precipitated an end to the coalition with the Socialists, then under chancellor Franz Vranitzky, and led to an early election in November in which the FPÖ doubled its vote. Haider emerged as the triumphant "media gladiator" in the new age of electronic populism. The Steger era left many members in the FPÖ with bad memories of coalition politics and with an aversion to any alliance with the Socialists. From now on the party followed a virulent opposition strategy aimed at creating a non-socialist Austria, free from the party book and the deadweight of ancient bureaucratic controls.

The party scored a series of successes in provincial elections and later bit into depressed industrial areas with rising unemployment. Many workers, especially in run-down nationalised industries, felt they had been left in the lurch by the Socialists. They warmed to the aggressive class-warfare language of the Freedom Party which stood up against the "penthouse trade unionists" with their fat salaries. The FPÖ also advanced in areas such as Burgenland where it had traditionally been weak and in 1987, for the first time in its history, gained a seat in the federal council (*Bundesrat*)[4] after elections in Vienna (see Table 3). The spate of spectacular victories continued with a grand slam in March 1989 in provincial elections in Tyrol, Salzburg and Carinthia (see Table 4). The coalition parties had tried to defuse the "Haider effect" by holding all three elections on the same day in the vain hope that the FPÖ leader would be overstretched. The tactic backfired and Haider continued with his relentless onslaught on former strongholds of the big two.

With the help of the ÖVP, Haider became governor in Carinthia and started to put through his policies of renewal to reduce the influence of the party book and privileges for politicians.

In 1986 Haider started out on a turbulent journey to reverse the fortunes of his party. He adopted controversial rhetoric particularly on immigration and neutrality and acquired some notoriety for ambiguous remarks on National Socialism. In June 1991 his analysis of the Nazis' employment policies cost him the governorship of the province of Carinthia. In 1993, one of the party's best-known politicians Heide Schmidt broke with Haider to form a Liberal Forum together with some colleagues in parliament. She had stood as the party's presidential candidate in 1992 but felt increasingly uncomfortable with its ideology. Schmidt criticised the party for populist demagogy on immigration and for switching to an anti-European course. In her opinion the party under Haider was becoming dangerously authoritarian and adopting illiberal policies which would lead to political isolation.

The FPÖ managed to survive these trials and tribulations and even won increased support at the polls. In the Vienna municipal election November 1991, the FPÖ played the immigration card and moved up to second place winning almost 23 percent of the vote. The ÖVP was beaten into third position and the Socialists suffered humiliating defeats in traditionally loyal working-class districts. The 1994 general election was a shock for both the ÖVP and the SPÖ who for the first time since 1945 could not muster between them a two-thirds majority in parliament necessary to pass important constitutional bills. Haider's FPÖ emerged as a medium sized party with the backing of a million voters and as one of the most successful right of centre forces in European politics.

The party was "psyched up" for electoral battle and 1998 set as the target date for moving in to the federal chancellory. At the beginning of 1995 the Freedom Party was restyled as a movement and adopted the title "*Die Freiheitlichen*," designed to fulfill the 1998 project and inaugurate a new "Third Republic." This radical programme of political renewal was to end the sleaze factor in the Austrian party state. It was generally interpreted as a policy of "total war" on the old parties. In the short term the greatest danger to the red/black coalition was not Haider but the SPÖ and the ÖVP themselves. Internal differences in the coalition precipitated a premature election at the end of 1995 which paradoxically rescued Vranitzky's Socialists and temporarily stemmed the successes of the *Freiheitlichen*. The advance was resumed however with the elections to the European parliament in October 1996 and the Vienna municipal elections held on the same day. The SPÖ lost its absolute majority in Vienna, suffering humiliating losses once more in its working-class strongholds to Haider. Haider's FPÖ had now become an important "power player" whose ambitions could not be ignored. What these ambitions were and what exactly the party was playing at was not so obvious. Many saw the FPÖ as sinister and dangerous although for others it was the joker in the pack. The party under Haider remained an enigma but a phenomenon which could not simply be dismissed.

FIGURE 1
PARTY COMPOSITION OF POST-WAR GOVERNMENTS

Year	Parties	Government Type	Period
1945	ÖVP/SPÖ/KPÖ	Provisional government	Occupation
1947		National government	
1955	ÖVP/SPÖ	Great coalition	
1966			Independence
	ÖVP	Single-party government	
1970	SPÖ	Minority government	
1971	SPÖ	Single-party government 'The Kreisky era'	
1983	SPÖ/FPÖ	Small coalition	
1986	SPÖ/ÖVP	Great coalition	

Year	SPÖ %Votes	SPÖ Seats	ÖVP %Votes	ÖVP Seats	(WdU)FPÖ %Votes	(WdU)FPÖ Seats	Greens %Votes	Greens Seats	Liberals %Votes	Liberals Seats	KPÖ %Votes	KPÖ Seats
1945	44.60	76	49.80	85	-	-	-	-	-	-	5.42	4
1949	38.71	67	44.03	77	11.67	16	-	-	-	-	5.08	5
1953	42.11	73	41.26	74	10.95	14	-	-	-	-	5.28	4
1956	43.05	74	45.96	82	6.52	6	-	-	-	-	4.42	3
1959	44.79	78	44.19	79	7.70	8	-	-	-	-	3.27	-
1962	44.00	76	45.43	81	7.04	8	-	-	-	-	3.04	-
1966	42.56	74	48.35	85	5.35	6	-	-	-	-	0.41	-
1970	48.42	81	44.69	78	5.52	6	-	-	-	-	0.98	-
1971	50.04	93	43.11	80	5.45	10	-	-	-	-	1.36	-
1975	50.42	93	42.95	80	5.41	10	-	-	-	-	1.19	-
1979	51.03	95	41.90	77	6.06	11	-	-	-	-	0.96	-
1983	47.65	90	43.22	81	4.98	12	3.29	-	-	-	0.66	-
1986	43.13	80	41.29	77	9.73	18	4.82	8	-	-	0.72	-
1990	42.78	80	32.06	60	16.64	33	4.78	10	-	-	0.55	-
1994	34.92	65	27.67	52	22.50	42	7.31	13	5.97	11	0.26	-
1995	38.06	71	28.29	53*	21.89	40*	4.81	9	5.51	10*	0.29	-

Source: Austrian Central Statistical Office
* In 1996, a Liberal member defected to the FPÖ. The FPÖ also gained a seat from the ÖVP in a by-election in October 1996, giving the ÖVP 52 seats, the FPÖ 42 and the Liberals 9.

TABLE 1

ELECTION RESULTS TO THE NATIONAL COUNCIL, 1945-1995

TABLE 2
NATIONAL COUNCIL ELECTIONS, 1949-1995, RESULTS FOR THE WdU (INDEPENDENTS) AND FPÖ

Year	Party Title	Party Leader	Votes	%Votes	Seats
1949	WdU	H. Kraus	489,273	11.67	16
1953	WdU	H. Kraus	472,866	10.95	14
1956	FPÖ	A. Reinthaller	283,749	6.52	6
1959	FPÖ	F. Peter	336,110	7.70	8
1962	FPÖ	F. Peter	313,895	7.04	8
1966	FPÖ	F. Peter	242,570	5.35	6
1970	FPÖ	F. Peter	253,425	5.52	6
1971	FPÖ	F. Peter	248,473	5.45	10
1975	FPÖ	F. Peter	249,444	5.41	10
1979	FPÖ	A. Götz	286,743	6.06	11
1983	FPÖ	N. Steger	241,789	4.98	12
1986	FPÖ	J. Haider	472,205	9.73	18
1990	FPÖ	J. Haider	782,648	16.64	33
1994	FPÖ	J. Haider	1,042,332	22.50	42
1995	FPÖ	J. Haider	1,060,175	21.89	40

Source: Ministry of Interior, Vienna

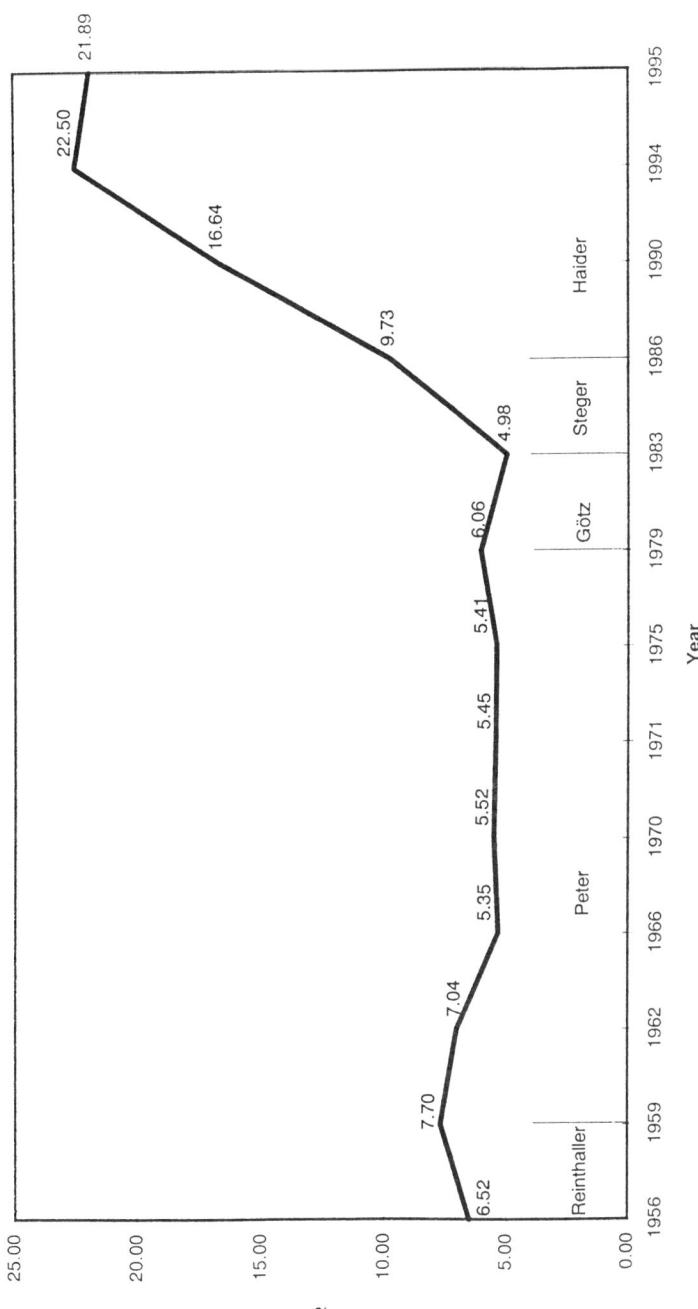

FIGURE 2
FPÖ SHARE OF THE VOTE IN ELECTIONS TO THE NATIONAL COUNCIL

TABLE 3

COMPOSITION OF THE FEDERAL COUNCIL, 1945-1996

Date	ÖVP	SPÖ	WdU/FPÖ	KPÖ
December 1945	27	23	0	0
December 1949	25	20	4	1
April 1953	25	21	3	1
October 1954	25	22	2	1
December 1954	25	23	2	0
November 1955	25	24	1	0
March 1957	26	24	0	0
July 1962	29	25	0	0
May 1964	28	26	0	0
November 1967	27	27	0	0
November 1969	26	28	0	0
March 1970	25	29	0	0
February 1972	28	30	0	0
November 1973	29	29	0	0
March 1982	33	32	0	0
March 1983	32	31	0	0
November 1983	33	30	0	0
December 1987	32	30	1	0
November 1988	31	30	2	0
April 1989	30	29	4	0
May 1989	30	28	5	0
October 1991	28	27	8	0
December 1991	27	26	10	0
March 1993	27	27	10	0
April 1994	27	26	11	0
October 1994	27	25	12	0
January 1996	26	25	13	0
October 1996	26	24	14	0

Source: Parliament Information Office.

TABLE 4
FPÖ PROVINCIAL ELECTION RESULTS, 1981-1996

PROVINCE	YEAR	PERCENTAGE
Burgenland	1982	3.0
	1987	7.3
	1991	9.8
	1996	14.6
Carinthia	1984	16.0
	1989	29.0
	1994	33.3
Lower Austria	1983	1.7
	1988	9.4
	1993	12.0
Salzburg	1984	8.7
	1989	16.4
	1994	19.5
Styria	1981	5.1
	1986	4.6
	1991	15.5
	1995	17.2
Tyrol	1984	6.0
	1989	15.6
	1994	16.1
Upper Austria	1985	5.0
	1991	17.7
Vienna	1983	5.4
	1987	9.7
	1991	22.6
	1996	27.9
Vorarlberg	1984	10.5
	1989	16.1
	1994	18.4

Source: Verbindungstelle der Bundesländer, Vienna.

NOTES

1. For an historical analysis of the Third camp, see L. Höbelt, *Kornblume und Kaiseradler* (Oldenbourg: Verlag für Geschichte und Politik, 1993), and M. Riedlsperger, *The Lingering Shadow of Nazism: the Austrian Independent Movement Since 1945* (New York: Columbia University Press, 1978).

2. A special feature of post-war Austria has been the development of social and economic partnership. This assumes good co-operation between the organisations representing the employers and employees and a talent for reaching an acceptable compromise. The principal forum is the parity commission on wages and prices which seeks a compromise solution to avert industrial conflict and strikes. The parity commission is a voluntary institution without any basis in law. It meets under the chairmanship of the chancellor with representatives from the government, the trade union federation, the chamber of labour, the federal economic chamber and the chambers of agriculture.

The commission has no legal means of enforcing recommendation. In practice the parity commission with its sub-committee for wages and prices and a special advisory committee for economic and social affairs has played a pivotal role in the management of the Austrian economy. The system is attacked by the FPÖ as a "shadow government without any democratic legitimacy."

The social partners are being forced more to justify their existence and are under increasing pressure to modernise their organisations.

For a full discussion of Austrian post-war politics, see K. Luther and W. Müller, eds., *Politics in Austria: Still a Case of Consociationalism?* (London: Cass, 1992); V. Lauber, ed., *Contemporary Austrian Politics* (Oxford: Westview Press, 1996); H. Dachs et al., eds., *Handbuch des politischen Systems Österreichs* (Vienna: Manz, 1997); W. Mantl, *Politik in Österreich* (Vienna: Böhlau, 1992).

Background information on politics in Austria and parties besides the FPÖ can be found in the appendices.

3. See M. Sully, *A Contemporary History of Austria* (London: Routledge, 1990).

4. There are two chambers of parliament: the national council (*Nationalrat*) and the federal council (*Bundesrat*). The former is elected every four years by the voters on the basis of proportional representation. The members of the federal council are elected by the provincial parliaments, their number being in proportion to the population of the individual provinces.

Draft bills pass through the national council and its committees, and are passed on to the federal council; this, in most cases, can only delay legislation since, if the national council once again puts forward its original bill, it is authenticated and passed as law. In only a few cases – and only then, when through a constitutional provision the competence of the provinces is restricted – has the federal council the right of approval and with this an absolute veto (article 44, paragraph 2 of the federal constitution; see also Appendix I, Figure A1). For more information on the federal council, see W. Zögernitz, "Reformen der Geschäftsordnung des Bundesrates," in H. Schambeck, ed., *Bundesstaat und Bundesrat in Österreich* (Vienna: Österreichische Staatsdruckerei, 1997), pp. 429-51. For a discussion of the legislative process, see H. Neisser, *Unsere Republik auf einen Blick* (Vienna: Ueberreuter, 1996).

2

ORGANISATION '95

A new organisational outfit was adopted at a party conference in Linz in January 1995. Whilst the electoral base of the party had expanded to include around a million voters, its membership had hovered at just over 40,000. Changes were considered necessary to reach the new circle of sympathisers which would provide information and eventually bring them firmly into the orbit of the organisation. For this purpose the party was to be remodeled on the lines of a citizens' alliance and a special "Alliance Office" (*Bündnisbüro*) was set up for elections and co-ordination work. The party was to become more of an electoral movement on the lines of American parties. It was to open up and be responsive to all segments of the population who could identify with some of its aims but who were not prepared to join as full members. The former socialist chancellor Bruno Kreisky had adopted a similar line in the 1970s when he called for voters to go "part of the way together" in an "open party."

At the special Linz conference, the Freedom Party of Austria also changed its name to simply *Die Freiheitlichen* dropping the words "party" and "Austria." Opponents took this as a sign that Haider was opposed to political parties per se and had little interest in Austria. The new-look F, as it became known by the media, pointed out that the Greens and Liberals made no mention of party or Austria in their titles and ridiculed the ÖVP and SPÖ as the "old parties," out of touch and out of date. The frequent use of words like "old" in this connection and "movement" conjured up sinister connections with Nazism for many looking to brand the F as rightist extremist. The Nazis too, they pointed out, were fond of talking of the "old" parties and their "movement." This seemed a little overstretched since the Greens had a tendency to refer to the SPÖ/ÖVP as "old" parties and there was also a Green "movement," a "peace movement," a "pan-European movement" not to mention a "labour movement."

By the summer of 1995 some nostalgia for the old FPÖ was apparent and Haider professed some boredom and irritation with the "F" title. The FPÖ was rehabilitated with the explanation that this had always been the core party and the *Freiheitlichen* was the wider electoral movement.

At the conference in Linz the party in parliament was given an enhanced role and took over responsibility for main policy statements from the general secretariat. The leader of the party in parliament, (the *Klubobmann*), and/or the party chairman, was to be the central spokesman for the party. He was to be assisted by the deputy leader of the parliamentary party.

The basic thinking behind the changes was that parties in the traditional sense were no longer appropriate for the modern age of electronic politics. A mass membership party such as the SPÖ had changed little in its organisation since the beginning of the century. The ÖVP had made several attempts to liberate itself from its archaic league structures which included a motley collection of farmers, public service employees, industrialists and working people. The FPÖ was not constricted by old party machines reluctant to adapt or surrender power. The leadership also saw that the time was ripe for changes, in the wake of electoral success, and it could reckon with little dissent.

Within the F hierarchy, expert panels were formed for specific policy areas to exchange information at federal and regional level. For this purpose an <u>information centre</u> was to be formed to collect data and cultivate contacts with members and supporters. A "hotline" was to be established to facilitate the dissemination of information. The centre was to ensure smooth co-ordination between the parliamentary party, the provincial parties and members and sympathisers. Information was to be ready on-line on current day-to-day issues. Appointments with members of parliament by ordinary members were to be arranged by the information centre. Proposals from politically interested citizens in the country were to be funneled through the centre and transmitted for consideration to the relevant department.

To regain some credibility for politicians and the profession of politics, deputies of the F in parliament and the provincial legislatures, as well as mayors and town councillors, were asked to accept limits on their income from public sources. Money earned in excess of around five times the average wage was to go to charity and be used for social work. Politicians' privileges and multi-pensions were burning issues; this gesture was to signal a new readiness to set an example and voluntarily accept a reduction in living standards which was demanded from ordinary people. This was dismissed as a cheap gag by the coalition parties, who suspected that ways round the rule would be found (see also Chapter 11).

THE CITIZENS' MOVEMENT 1998

The "Citizens' Movement" (see Figure 3) was targeted at mobilisation and recruitment of support for the next election scheduled for 1998. It sought to promote contacts with citizens interested in a dialogue with the party and to offer them the opportunity to take some part in the movement's internal decision-making process. It consisted of concentric circles including an inner membership core, an "active-circle" and an outer "info-circle." Those who wished to participate in the scheme were to abide by the democratic system and the Austrian republic and to reject the use of violence.

The "info-circle" was open to those over the age of 15 who agreed with these principles and paid a small fee. They could belong to another party and did not need to be of Austrian citizenship but would still be entitled to receive an info-card. This would give them access to a "hotline" telephone connection and enable them to make suggestions and recommendations to the party. Information would also be readily available on parliamentary and political work.

The "active-circle" was possible only for Austrian citizens over 18, who were not already members of another party and who also paid a fee and subscribed to the above basic principles. Those in this group could take part in seminars and working groups of the party and likewise had access to the hotline service. Card holders in this category were guaranteed a reply to a request from a parliamentary deputy within two weeks. They could also take part in electoral conventions and decide on candidates to contest elections. To be on the party list required the support of 50 party members or the support of 200 of those entitled to vote in a particular regional constituency. This had to be secured at least two weeks before the convention.

The core of these concentric circles was formed by a "party-circle." Members in this group were to agree to the ideas outlined in the programme of the party. They had automatic access to the other circles in the party and could take part in the decision-making process. Their duties and rights were already anchored in the party's statutes concerning organisation. They had all the rights enjoyed by those in the outer circles and were guaranteed a reply from a member of parliament to a query within a week. Finally, there was the possibility of the more expensive "promoters' card" which included the active-card package and offered, in addition, invitations to special events and social occasions.

Each category of affiliation issued plastic cards to supporters or members. Little dissent was voiced at the conference held in the ultra-modern Linz Design Centre. Originally the idea had been to adopt the English name "card" for info card, active card etc. but this was rejected and substituted with the German word "*Karte*." Delegates warned against caving in to creeping Americanisation and argued in favour of German to stop the rot and the spread of the "American Way of Life." Haider voted in favour of the offending word "card" but to no avail and conference remained true to the German language. A "hotline" that was to be set up escaped the purge of the purists either through oversight or translation problems.

An <u>electoral convention</u> was to democratise the proceedings for the selection of candidates for elections. Party members and active card holders would be able to take part in the process of candidate selection.

The electoral campaign of the party would be mobilised in the constituencies long before the rival party machines had swung into action. It was intended to get the campaign off the ground in each of the 43 constituencies about two months before an election. In all, around 20,000 people were to be deployed in the electoral exercise. The thinking behind this was to generate momentum in the country at large which would create a bandwagon effect, comparable to the impact of primaries in the early phase of an American presidential election. The electoral conventions were conceived to be a vital springboard for any successful electoral battle of the party. They were not operative in the snap election of 1995.

Party members and active-card holders could apply for a ticket to attend a convention in their area about six weeks before an election. Every party member had the possibility to be selected as a parliamentary candidate. The active-card holders, if backed by 200 voters in the constituency or by 50 members, could also put their names forward to go onto the party list. The final selection would be made by elections at the convention. The '98 project aimed to involve citizens and give them some influence in shaping the politics of the party, free from the constraints of anonymous institutions, chambers and trade unions so dominant in the party state. From now on the "F movement" was to be a citizens' movement of allied sympathisers as well as full members.

CONTRACT WITH AUSTRIA

The leadership, unlike the rank and file, did not shrink from importing a trans-Atlantic political culture and "new wave" ideas. At the Linz conference the project for a "Contract with Austria" based on the Newt Gingrich model was launched.

At a rally for a "Contract with Austria" in Vienna in June 1995, Haider elaborated his thesis that Austria was an "underdeveloped democracy," in comparison with other west European countries, and was badly in need of renewal and regeneration. He deplored the limitation on the possibilities for citizens to shape their own destiny and called for more direct democracy and freedom of opinion, free from state monopolies such as radio and television.

According to Haider, Austria was a country ruled by powerful bureaucracies and parties in which individuals were deliberately kept in dependence. Abuse of power had brought with it corruption and misappropriation of public funds, so that essentials such as social and health services were in serious financial difficulties. The next generation would have to pick up the bill unless something was done urgently.

Article 1 of the Austrian constitution states that "Austria is a democratic republic. Its law emanates from the people." For Haider this had become a charade and it was time to sort out the discrepancy and to enable citizens to share in the decision-making process. He restated his belief that representative democracy had become an empty phrase; parliament had been hijacked by party machines. People had become alienated from remote and unaccountable elites and felt estranged from politicians and government.

Haider's contract with Austria aimed at creating an open, living democracy with active participation in which people would be freed from party political tutelage. It served also to involve experts in policy formulation for a possible participation in government. Unlike the two main parties the F did not have the luxury of drawing on a body of experts from the social partners. Think tanks in the Anglo-American sense were in any case in Austria something of a rarity.

Following Kreisky's example, Haider engaged around "1,000 experts" to work on this project divided into 8 main working groups for special subject areas (see Figure 4). They were to be given a free hand to work out their recommendations but had to be in agreement with the principles in the party programme. The experts were entrusted to work out a contract for the renewal of Austria and for a constitutional and democratic reform programme which would pave the way for a prototype citizens' republic.

FIGURE 3
CITIZENS' MOVEMENT '98

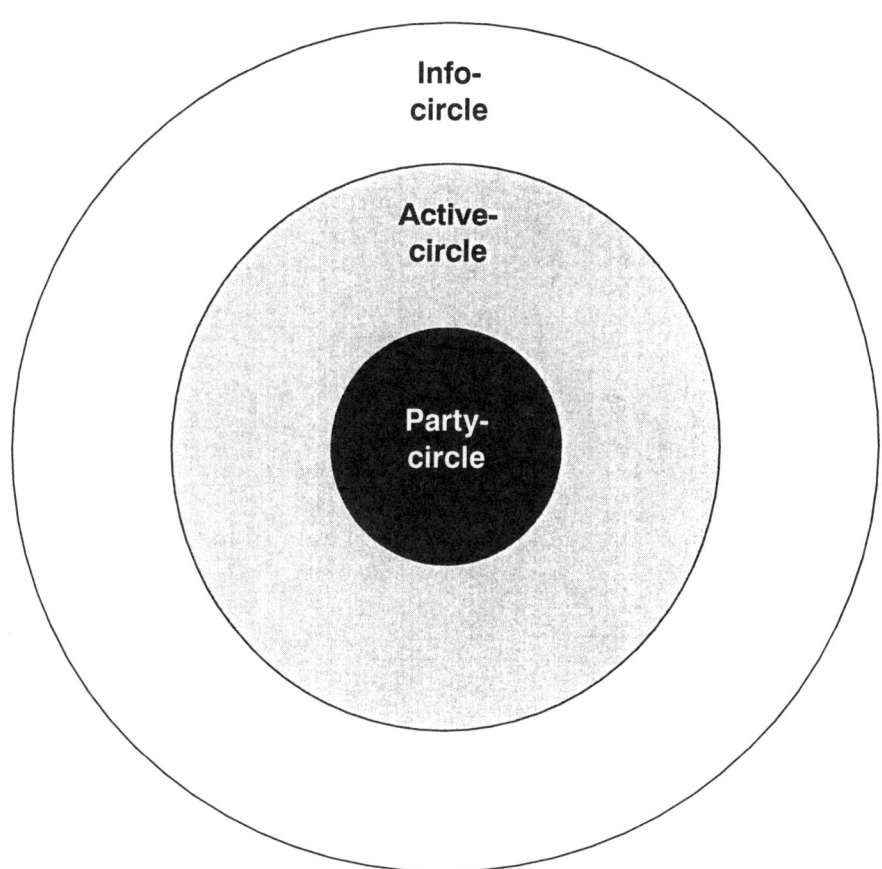

In addition, a Promoters' card is available.

Source: FPÖ.

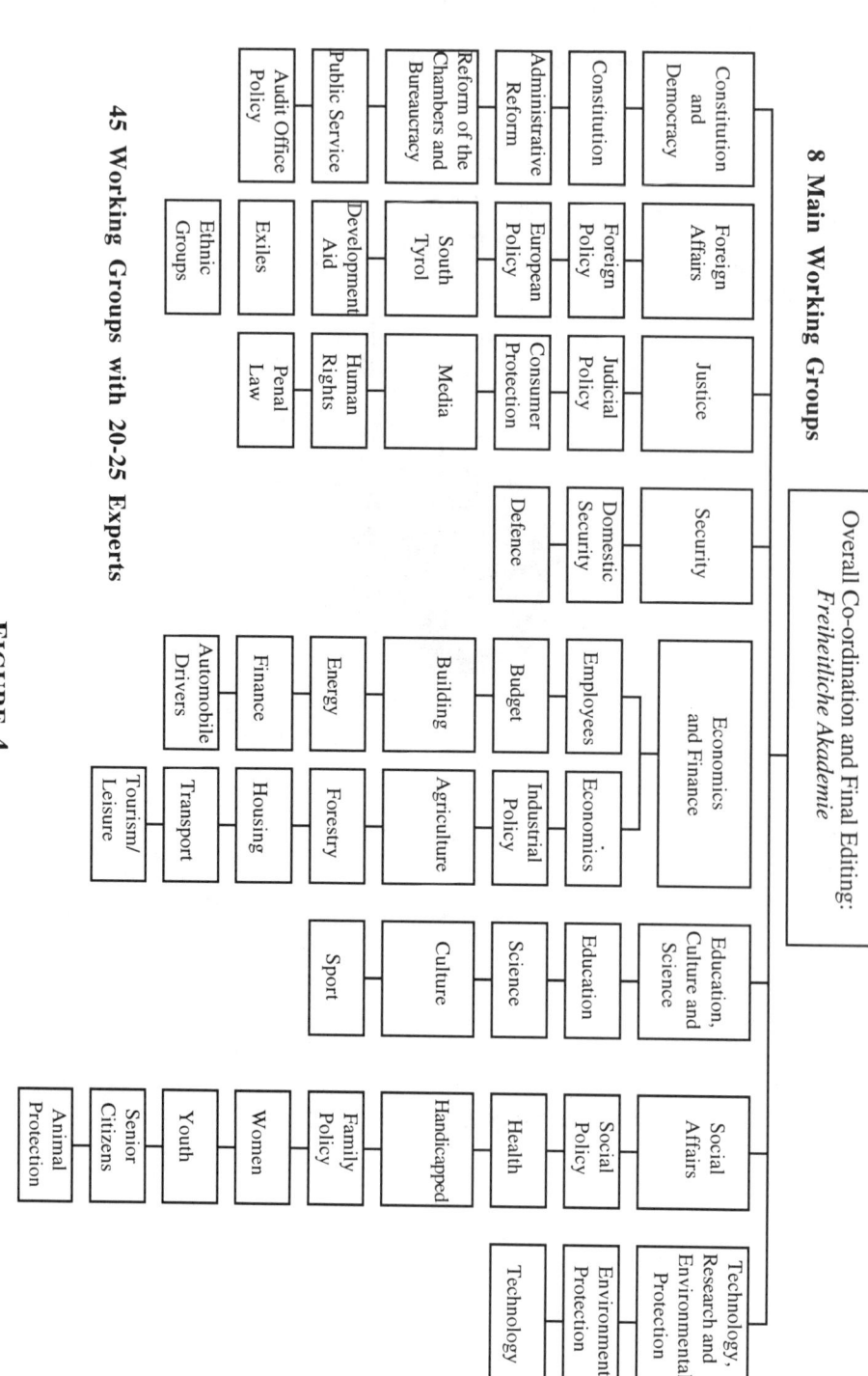

Source: FPÖ

**FIGURE 4
CONTRACT WITH AUSTRIA**

3

THE "THIRD REPUBLIC"

The Contract with Austria was to take the country into a "Third Republic" built on the model of a citizens' republic. In 1945 a Second Republic was created dominated by the two main parties with a high level of consensus and stability. Party patronage, in the form of a highly sophisticated *Proporz*, served to mobilise support for the two political encampments and solved the crisis of distribution which had plagued and undermined the First Republic (1918-34). The Second Republic was characterised by elite accommodation and passive participation by the people. Haider and the FPÖ paid homage to the "success story of the Second Republic" but maintained that its darker side of political corruption and immobilism had to be tackled. Sclerotic institutions and an archaic bureaucracy underpinned big corporate interests, leaving little scope for individual freedom. The Third Republic was to end the "dead-end democracy" and breathe new life into a stale political system.

This analysis had much in common with academic critiques of the malaise in the Second Republic. According to the political scientist Anton Pelinka (close to the SPÖ) Austria was a "late developer" and democracy, when it arrived, came on the "bayonets of the victorious powers."[1] It was both a highly politicised society and centrally controlled by the party machines but at the same time an apolitical society where active participation by the population was neither desired nor sought. It was government for the people but not by the people. Historically this could be explained as in Austria both of the main parties and interest groups had predated a mature, functioning parliamentary system. As Pelinka pointed out, life in Austria centred around the two big political parties – "whoever wants anything in the Second Republic must go for it through the SPÖ or the ÖVP." They were the "gate-keepers" of the post-war political system. Working through, by, or in the parties ensured appointments to the constitutional court, management in nationalised industries and careers and jobs in the schools. This had created the same kind of dependency-culture as in other western European systems but it was party, rather than state, directed. Pelinka recalls the case of the Olympic village in Innsbruck where both the SPÖ and the ÖVP had a higher voter to membership ratio than elsewhere in Innsbruck. "The reason has certainly nothing to do with the special political activity of the inhabitants of the Olympic village but has more to do with political clientelism: Whoever wants a publicly subsidised apartment feels obliged to join the protecting association of a big party." Pelinka calls this a type of politics which administers opium from the top downwards.[2]

The party/chamber state peaked in the late 1970s and then ran into trouble with the "greening" and secularisation of Austria. A liberation process from the neo-corporatist party/chamber state started which heralded the end of the Second Republic in its post-war form. Some writers and ÖVP politicians had spoken of the beginnings of a "Third Republic" long before Haider. When the phrase was adopted into the "F" terminology it was no longer regarded as "politically correct."

The ÖVP in Styria had come up with the idea of a Third Republic in the mid-1980s. In this republic the People's Party was to play a leading role in a government of the best minds, i.e., "an all-Austrian government." This would have opened up the way for the party, then in opposition, to the cabinet and the SPÖ and FPÖ showed little enthusiasm for the concept. One of those most concerned with this project in Styria for the ÖVP was Bernd Schilcher who formulated his ideas in *Die Furche*, 13 February 1985. Schilcher shared much of the criticism of the FPÖ of the Austrian party and chamber state. In an article which appeared in the 1996 Yearbook of the Freedom Party, Schilcher updated some of his ideas.[3] In this contribution, Schilcher criticised the widespread "pathological" rejection of Haider which had distorted a clear appreciation of some fundamental reforms necessary for Austria. He also rebuked Haider for his "Narcissism" rather than his "Nazism." Schilcher supported wide-ranging reforms such as the introduction of a new electoral system on the English model, decentralisation, increased citizen participation and an upgrade in the status of the president. Schilcher saw this as a means of moving away from the one-dimensional fixation with government for the business of state. He concluded that it was irrelevant whether this concept was called a Third Republic, since what mattered most was devising new rules of the game for the conduct of the state.

Ten years even before the ÖVP model, Alexander Vodopivec, a political historian, had written a book on The Third Republic – Power Structures in Austria. His theory was that constitutional practice and theory were at odds and that the 1920/9 constitution existed on paper but not in practice. He listed numerous examples of discrepancies and came to the conclusion that Austria was in fact well on the road to the "Third Republic." He spoke of the Austrian Trade Union Federation as a "state within a state" and referred to the fairy tale of freedom of opinion in the Alpine republic.

For Haider it was clear that "all was not well" with the Second Republic. An age of uncertainty confronted its people who were unsettled by the lack of leadership offered by politicians and church leaders alike. Many questions were awaiting answers and many problems were left in limbo unsolved. As Haider said in his speech on the Contract with Austria in June 1995, "We have in effect got into a vicious circle through these misguided policies. We have the oldest students, the youngest pensioners, the highest rate of divorce and the lowest birth rate, more holidays and the least working days, the fewest industrial workers, and the most bureaucrats.... It can't go on like this, we need changes in this country which give rights to the citizens and not the parties."

The concept of the Third Republic was understood by the *Freiheitlichen* as a kind of dialectical development of the existing system. All the other four parties in parliament interpreted it as a declaration of war on the status quo and suspected a take-over bid backed by sinister motives. They were not reassured by the title "Third Republic" which evoked unfortunate associations with the Nazi Third Reich. The F argued that the red/black *Proporz* democracy had created another set-up than envisaged under the constitution. Its warming up of the idea of a Third Republic was to restore some harmony between constitutional theory and practice. Like many items in the FPÖ repertory, it was not original. The Third Republic was in effect a slogan used by different writers to signal that an era in post-war politics had come to an end.

"THE COUNTRY MUST CHANGE"

According to the Freedomites the country was already going through a stage of fundamental change but the "old parties" had either not woken up to this or did not want to see what was happening.

The model for the Third Republic was mooted in a book, *Weil das Land sich ändern muß!* (The Country Must Change), published by the educational and research department of the party

(*Freiheitliches Bildungswerk*) in 1994. It was written by academics and politicians and amounted to a critique of the political, economic and cultural Establishment. More than this it suggested an alternative way forward and not surprisingly provoked an emotional response from those entrenched in power who stood to lose from the renewal.

This reform model was to address Austrian specialties such as social and economic partnership with its parity commission for wages and prices, powerful bodies without any democratic accountability or legitimacy. These institutions had evolved in the Second Republic to become a kind of shadow government but were not even mentioned in the Austrian constitution. The authors of *Weil das Land sich ändern muß!* were also critical of the monopoly in the state broadcasting service (ORF) which they believed led to a distortion of information comparable to the manipulated propaganda in the former Soviet satellites.

Concrete proposals for constitutional change included a new role for the federal president. He was to chair cabinet meetings and lay down guidelines for the entire government, in addition to his existing titular functions as head of state. With this upgrade, the authors considered that it was possible to dispense with the office of the federal chancellor. This reform was too radical for many particularly on the left who feared a presidential republic would develop based on the notion of a "strong man." The president was, as in the past, to be directly elected by the people and would therefore enjoy considerable democratic legitimacy. He, together with the cabinet ministers, would form the government which would be considerably streamlined as compared with the past. "Lean government" on the Swiss model was the aim. Precise details for the whole system were not explained in this working hypothesis which was intended to serve as a basis for further discussion.

The Second Republic established in 1945 adopted the 1920 constitution of the First Republic with the amendments of 1929.[4] Constitutional experts differ on the real powers the president could realistically exercise. An "active" president could interpret the powers given to him more widely without constitutional changes. The amendment of 1929 implanted a presidential element into the parliamentary system. This was typified by a directly elected president who formally had powers to appoint and dismiss the chancellor and other members of the federal government. For much of the life of the Second Republic the post caused little controversy and fitted harmoniously into the political duopoly. A "red" president balanced a "black" chancellor and for the most part played along with the norms of the great coalition (1945-66). Even when formally the coalition fell in 1966 the old red/black mentality persisted. Matters changed with the election of Kurt Waldheim as president in 1986. Waldheim, a candidate of the ÖVP, was the first president of the Second Republic who had not been put up by the Socialists. His wartime past in Hitler's army and his apparent selective amnesia on this episode in his career caused international outrage. He was placed by the United States on the "watch list" which barred him from entering the country as a private citizen. The ensuing uproar in Austria led many to rethink whether the office of federal president was worth the trouble. Some Greens proposed outright abolition of the post and even liberal conservatives like the constitutional expert Manfried Welan thought Austria could get by without a president.[5] The federal president, according to Welan, contained "post-imperial" elements on the one hand and "pre-dictatorial" on the other. The office arose out of conflict not consensus in the period after the First World War and had now largely lost its relevance. The presidency evoked expectations of harmony and unity which could not be fulfilled and which could be dangerous for a democratic republic. Welan concluded that Austria had long since ceased to be a monarchy and had no need of a "substitute Kaiser." Arguably after 1945, the president fulfilled a bridge function for a republic in transition from an authoritarian-totalitarian system to a free democracy. In the 1990s, Austria needed no crisis manager or *Führer* figure. It had come of age; the days of paternalism were over.

Waldheim came and went and was succeeded by Thomas Klestil in 1992, an internationally respected diplomat with good connections in America. He was to be an "active president" and aimed to make greater use of the powers entrusted in the post by the constitution. He took his election slogan – "Power needs control" – seriously and like Haider was concerned with the influence of ossified and outmoded political institutions. Klestil was the candidate of the conservative People's Party and so continued to be a thorn in the flesh of those on the left who sought to reduce the office to a chiefly ceremonial post. As a diplomat Klestil showed ambitions in the field of foreign affairs and irritated the SPÖ by aspiring to represent Austria abroad. At the end of 1994 Klestil gave an interview stating that Austria's future path led to the West European Union (WEU) and the NATO Partnership for Peace. This had constitutional implications and affected the neutral status of the country. For the SPÖ Klestil had questioned the national consensus on security policy in an inappropriate manner. Josef Cap, then SPÖ party manager, suggested it was high time to look at the constitutional "grey areas" of the powers of the federal president. He regarded some of these as monarchistic. He was supported in this by the Liberal Friedhelm Frischenschlager. (Only eighteen months later Cap himself gave a controversial interview in which he opened up a discussion on WEU and NATO membership for Austria.) The president was supported on this occasion by Haider and the People's Party who fought off attempts to clip the wings of the office. The status quo was preserved with the help of the Freedomites against a foray from the left of centre "traffic lights" coalition including the SPÖ, Greens, and Liberals.

The Third Republic idea of the FPÖ had poured oil on troubled waters by suggesting concrete, radical change. The party's opponents criticised it for creating a destructive influence and destabilising the political system. However, many, like the ÖVP second president in the National Council Heinrich Neisser, agreed it was time to rethink the triangular power relationship between president, parliament, and government.[6]

Haider always liked to think that his party set things in motion in Austria and that on many questions the other parties acted in response to and not in advance of the Freedom Party. This was as true for immigration as for the question of privileges enjoyed by politicians. In some ways, however, the FPÖ managed to shore up resistance to change and an open way of looking at constitutional reform. A Freedom Party proposal, however intelligent, was destined to cement opposition from the "traffic lights."

Other proposals for reform put forward in the pamphlet "The Country Must Change" included giving more powers to the second chamber (the *Bundesrat*) to represent the interests of the provinces more effectively. Haider was in favour of an upgrade for the second chamber on the lines of the U.S. Senate. He saw the *Bundesrat* as an unjustifiable expense in its current form. The party also wanted to devolve power where possible and supported the principle of subsidiarity. Laws should be formulated so that they could be understood by the average citizen. In addition a catalogue of basic rights for citizens was floated. These were to include a right to one's native country (*Heimat*) and the right to a cultural identity of the individual and of ethnic groups. There was to be a right to expect efficient, thrifty and intelligent use of public funds. Other rights included protection for the family, provision in old age and the right to an intact environment. These were to be enshrined in a kind of "Magna Carta." Citizens were not only entitled to rights but were also expected to fulfill duties. Possible duties were to include military service irrespective of gender and an obligation to exercise solidarity, e.g., for the elderly and unemployed.

A central element in the new scheme was the extension of direct democracy on the lines of the Swiss model. The authors of *Weil das Land sich ändern muß!* wanted to see greater use of plebiscites and less bureaucratic obstacles. The Freedom Party thinking was based on the notion that the average voter was adult enough to give a reasoned opinion on any issue, but that governments tended to relegate instruments such as the consultative referendum. The call for more direct democracy had often been raised in the past and some reforms had been introduced. It was generally conceived as an advance in democratic thinking until the "F movement" latched on to the scheme. This was enough to kill any productive debate. By now initiatives of the Freedom Party were automatically rejected as a matter of course by the other parties and denounced alternatively as reactionary and/or dictatorial.

The fear was that the combination of a stronger president with greater direct democracy would lead to a downgrading of parliament. By way of reply some FPÖ members such as Herbert Haupt, one time third president of the National Council, believed that plebiscites would have to be used economically. Once again details of how the Third Republic would work in practice had been tantalisingly omitted. It amounted to an intellectual self-assembly kit in which the reader was invited to engage in amateur constitutional "Lego" building. In effect the vagueness of the Third Republic was its salvation. It could be changed at whim and most things could be reinvented if necessary. Even the title could be dropped, if convenient. In any case there was no real link between a plebiscitary democracy and a potentially impotent parliament. Political scientists had long noted that in Austria politics was lopsided in favour of representative democracy. The plebiscitary element was underdeveloped and overshadowed by a system in which the parties and the economic chambers pulled the strings. Referenda in the Second

Republic were rare and had hardly come about in response to people power. In 1978 Chancellor Kreisky asked the people to vote on the commissioning of a nuclear power plant in the hope of a positive response and to get out of a tight corner. To his chagrin the electorate narrowly defied the recommendation of his government and the social partners.

The procedure for a referendum is controlled by parliament, blocking initiative from voters. This leads to the paradox that the people are dependent on the main organ of representative democracy (parliament) before a referendum can take place. There are few possibilities to initiate a referendum against the wishes of the majority in parliament. Pelinka has called this an "overcontrol" by parliamentarism and the party state. The instrument of direct democracy should not be used, according to these rules, against the explicit wishes of these controllers.

The other referendum in the life of the Second Republic was an obligatory one on the question of the EU membership and the constitutional changes this involved in 1994. Popular initiatives were similarly under-used and unheard of before 1964 just before the Great Coalition fell in 1966. The FPÖ call for more plebiscites à la Third Republic provoked the criticism that this was leveled against parliamentarism, since the government could seek legitimacy not from parliament but from the "people." According to Professor Pelinka, the real danger with a plebiscitary trend for a democracy is that it can awaken too many expectations. This can lead to disillusionment with democracy per se and pave the way for Caesarism or Bonapartism. However, Pelinka adds that there is also a chance that parliament could become a livelier body and emancipate itself from the shackles of the party state.[7]

The pamphlet of the *Bildungswerk* to renew Austria also included familiar Freedomite calls to abolish compulsory membership in the professional chambers and to open up the system of social partnership to make it more transparent and responsive to the general public. This call was no monopoly of the FPÖ and reflected a general dissatisfaction with the persistence of semi-feudal institutions and decision-making procedures.

In addition the authors wanted to roll back the influence of parties in nominating positions in schools, supervisory boards and the constitutional and administrative courts, etc. They considered that membership of a political party was incompatible with some posts such as university lecturers, judges, army officers, diplomats and state attorneys. They wanted to take out participation by the political parties in the running of banks, insurance companies and building societies.

Party membership although declining has traditionally been high in Austria. Around 14 percent of Austrians still belong to a political party and the magic "party book" has often been the passport to a lucrative job, an apartment or credit from the bank. Party political intervention has been one of the less attractive features of the Austrian post-war success story. Many concede the need for reform but are also dependent on the system. The Freedom movement argued that, unlike the big two parties, it was not enmeshed in this network of party power and therefore was in a position to genuinely change the status quo. Neo-conservatives in western Europe and America have since the 1980s called for a reduction in state intervention in the economy and in the lives of individuals. In Austria because of the peculiarities of the political system this was coupled with the call to reduce the power of political parties. In addition the F movement supported the concepts of limited state intervention and maximum individual freedom. The state was to be entrusted with internal and external security, efficient management, rule-making (implementation and adjudication), basic social welfare provision, education, culture, and currency and fiscal policies. The Austrian system was considered to be overloaded and over-regulated. Some features were reminiscent of the former Eastern bloc in the opinion of the *Freiheitlichen*. They often point out that there are more public service employees than industrial workers and that party political wheeling and dealing has stilted the growth of a free market economy. In some areas, such as social security, pensions and health, this inbuilt inefficiency has rendered the system dysfunctional. Waste and mismanagement have led to an alarming budget deficit. The Freedom Party calls for drastic cuts in public expenditure and a real privatisation offensive to include not just banks and insurance companies but also the Austrian Airlines, railways and postal services as well as the broadcasting network.[8]

In the opinion of the FPÖ there were unacceptably high levels of party political intervention in radio and television. The party frequently complained that its meetings were not given fair coverage and sometimes not reported at all. The party complained that the ORF was in effect a government mouthpiece. A long-standing FPÖ demand was for an end to the monopoly of the Austrian broadcasting network (ORF). In 1995 an effort was made to end the state monopoly in radio and television. Franchises were issued to private radio operators and after some hitches the first radio station independent of the ORF went on the air in September 1995. Much to the chagrin of the FPÖ, the television monopoly remained until a fresh effort at liberalisation was made in 1997.

In addition the party called for an end to press subsidies given by the government to newspapers. The daily newspaper *Die Presse* received 38.7 million Schillings from the state and the rival quality daily, *Der Standard*, received a public subsidy of 33.7 million Schillings in 1996. A provincial newspaper in Graz, *Neue Zeit*, with a readership of only 100,000 got 38.2 million Schillings. The Freedom Party questioned the way in which this money was handed out and claimed it promoted a press loyal to the government. The government replied that without such assistance the quality papers would go to the wall. The FPÖ believed this system was a distortion of competition and took its case to the European Court. This argument was rejected with the reasoning that the subsidies were only given to Austrian concerns and did not discriminate against other EU member states.

The "Third Republic" provoked a storm of controversy especially after the FPÖ gains in the 1994 election. The main accusation, especially from the left Social Democrats, Greens and Liberals, was that it would lead to an authoritarian presidential or even *Führer* republic in which parliament would be politically castrated. To some extent this reaction can be understood in terms of a ritualistic exchange of demagogy between rival sides. Not surprisingly the condemnation of the "F" under Haider became more vitriolic with electoral success. Outright rejection of the ideas of the Freedom Party had become an automatic genuflection for many in the so-called left of centre "traffic lights" coalition. Most parties acknowledged the need for constitutional reform especially after Austria joined the European Union in 1995. Many accepted the case for modernisation in the political system, parties and the economy. It was a weakness of the opponents of Haider that they spent so much time lambasting his ideas that they failed to effectively project an alternative scheme. It was perhaps easier to engage in such a discourse than foil the FPÖ with a counter-strategy for renewal. In the SPÖ there had long been a deficit in intellectual thought compared with the brilliance of the inter-war period. Neither the Socialists nor the ÖVP were capable of reforming their own organisations to suit the modern age, let alone able to pioneer a project to change the country. The Liberals often lacked profile and punch and tagged along with the coalition. The ÖVP had invested much energy in working out a traditional party programme which for the most part was ignored even by its own people. The left Greens were often the only force which could effectively play the F movement at its own game. Through their tactics, political debate was not raised to a higher level but degenerated into a shouting match. The beneficiary of this was usually Haider or Green fundamentalists who also relished polarisation.

The concept of a Third Republic developed its own momentum after the 1994 election. The research department of the FPÖ fleshed out some ideas discussed in early pamphlets and these themes were consistently taken up by Haider in major speeches. "From party state to a citizens' democracy" was the slogan of the Third Republic outlined in a party brochure. Its main points were:

- Direct election of not just the federal president but also the provincial governors and mayors. People not parties were to have more say "on the French and American models." Some critics of Haider feared that directly elected representatives would mean more populism and feared that the parliamentary character of the system would be diluted. Some ÖVP politicians were in favour of more direct democracy and even in 1992 the provincial governor of Lower Austria, Siegfried Ludwig, had supported direct elections for the federal chancellor and for provincial governors. The Social Democratic provincial governor of Burgenland, Karl Stix, also supported this idea for provincial governors.[9]
- Abolition of the post of federal chancellor: "as in other western democracies, one head of state should be enough for Austria."
- The directly-elected president should chair the session of government "as in the USA."
- The government should be reduced to only seven members on "the Swiss model." These were to include: foreign affairs, justice, security (internal and external defence), social affairs, economics and finance, education, culture and science, and a ministry for technology, research and the environment. Originally in this rationalisation plan, Haider had suggested that the defence and

interior ministries should be merged. In November 1995 he adjusted this and preferred a combination of the interior ministry with the ministry of justice. Responsibility for the country's defences would come under the foreign ministry. The original idea of creating a kind of "super police and military ministry" had been compared by the SPÖ and Greens to dictatorial regimes. The FPÖ leader was accused of showing a "catastrophic lack of historical sensitivity." When asked why he had suddenly changed his mind, Haider casually replied that he had been convinced by the force of these arguments.

- Members of the government were to be elected by parliament after "hearings."
- The federal council (*Bundesrat*) should be able to exercise an absolute veto on legislation coming from the national council. An arbitration committee of both chambers should examine cases of conflict.
- The constitution should be changed to give the provinces and local councils more power and rights.
- Greater personality element in elections and the possibility of "ticket-splitting."
- Greater weight to be given to citizens' initiatives and petitions.
- Less bureaucracy and more scope for individual initiative.

In his book, *The Freedom I Mean*, Haider summed up the case for the Third Republic: "Post-war Austria exhibits authoritarian tendencies with a structurally-induced lack of freedom. It is dominated by parties whose representatives obediently follow their master's voice." This era, according to Haider, was coming to an end.

> We want to change this system and we make no secret of it.... We want to get rid of the corporate elements in this system and abolish privilege and corruption. We want to replace it with an open, democratic society of free citizens. We want to achieve this through democratic dialogue and debate.... What we have in mind is more than a change of government. We want to bring about a cultural revolution by democratic means. We want to overthrow the ruling political class and the intellectual caste."

For the F movement, culture, education, the universities and the arts were in the clutch of the '68 generation with its fancy ideas on ideologically-motivated multi-culturalism. Its so-called "progressive" experiments had led to a nihilistic society, the generation X, and the erosion of family values and solidarity. Ideological fantasies in art and culture had led to a deadening spiritual conformity. The F movement wanted to see a return to thrift, hard work, and family values.

The F leader's analysis of *fin de siècle* Vienna, not surprisingly, alarmed the ruling and intellectual elites who accused Haider of seeking to destabilise and destroy the Second Republic. In their opinion he was the gravedigger of political renewal and not its architect. Haider countered such attacks by saying that such fundamental constitutional changes would in any case require a two-thirds majority in parliament. He justified the Third Republic project with an unequivocal endorsement of the concept of freedom:

> Our freedom rules out any form of authoritarian domination. Our freedom rules out every sort of patronage which stops people from taking their own decisions. Our freedom rules out violence in politics because violence rules out intellectual competition. Our freedom rules out bureaucratic coercion because it stamps out individual and social initiative. Our freedom however demands order, results, responsibility, and a capacity to learn. To sum up – what we want is:
> - as much state as is necessary and as much freedom as possible.
> - no state-run human beings but a humane state.
> - less power for functionaries and more rights for citizens.
> - less squandering of public funds and more money for the private purse.
> - less politicised ethics and more ethical politics.
> - less group egoism and more concern for the whole community.
> - less oppression by the party machines and more freedom for citizens.
> - less subsidies for the lazy, more support for the industrious.
> - a break with bureaucratic chains and more intellectual autonomy instead.
> - an end to abuses on immigration and more rights for one's own native country.[10]

NOTES

1. A. Pelinka, *Zur österreichischen Identität* (Vienna: Ueberreuther, 1990), pp. 24 and 37.
2. A. Pelinka, *Windstille* (Vienna: Medusa, 1985), p. 155.
3. B. Schilcher, "Ein österreichisches Modell der III. Republik," in *Freiheit und Verantwortung* (Vienna: Freiheitliche Akademie, 1996), pp. 59-74.
4. For an examination of the role of the federal president, see M. Welan, *Das österreichische Staatsoberhaupt* (Vienna: Verlag für Geschichte und Politik, 1986). After an election the federal president usually appoints as chancellor the leader of the largest party in parliament. This assumes the election has produced a clear result. In other situations such as the formation of a minority government, the president can play an important role.
5. M. Welan, *Der Bundespräsident* (Vienna: Böhlau, 1992), pp. 101ff. See the interview with Waldheim in this volume for his views on Haider. For a discussion of Austria and its past in the wake of the Waldheim affair, see W.E. Wright, ed., *Austria, 1938-1988. Anschluß and Fifty Years* (Riverside: Ariadne, 1995).
6. Cited in *Die Presse*, 14 February 1994.
7. A. Pelinka, *EU-Referendum* (Vienna: Signum, 1994), pp. 19-20.
8. See T. Prinzhorn, *Wirtschafts- und Industriestandort Österreich* (Vienna: FPÖ, January 1997).
9. For potential constitutional problems in connection with the direct elections of provincial governors, see F. Koja, "Finger weg von der Direktwahl von Landeshauptmann und Kanzler!" *Die Presse*, 4 December 1995. Support for the idea was made by Karl Stix in *Die Presse*, 24 June 1996.
10. Quotes from J. Haider, *The Freedom I Mean* (New York: Swan, 1995), pp. 59-66.

4

"IDEOLOGY"

The Third Republic which sought to renew Austria drew on similar thinking current in America and Britain. It could be described as part Austro-Gingrich plus a dose of Helvetian democracy. It was in many ways a melange or less generously put, "a dog's dinner." Its advantage and appeal was its flexibility. As Haider once flippantly said, "call it what you will, Third Republic or New Austria – it's only a working title."

Like the old party machines the days of rigid ideology were over. Heide Schmidt had acknowledged this when asked where her Liberal Forum stood. She rejected the right-left political classification as inappropriate to the modern world. Political issues were important which could cut across the traditional ideological pigeonholes. Haider, too, believed ideology was finished; it had been responsible for some of the greatest catastrophes of the twentieth century.

When Haider was asked whether his movement was on the right, he replied, "We're neither right nor left, we're just in front." Haider himself was often described as a political chameleon who changed his ideas as often as his wardrobe. This genius for adaptation is part of the politician's make-up but it bothered his critics who wanted to know where he really stood. Some drew their own conclusions and labeled the Freedom Party, under its leader, as "extreme right." Definitions of this phrase are legion and may include several prerequisites, so that allocation of a political group to this category can be subjective and inconclusive. The negative connotations of the term mean that it can be used as abuse rather than a tool of analysis. The new political populist movements in the 1990s often defied neat definition since they responded readily to specific events.[1]

To the irritation of the left, Haider frequently cited famous Social Democrats who had influenced his politics such as Victor Adler, who had founded the movement in the nineteenth century, and Bruno Kreisky. He combined Popper's "open society" with Kreisky's "open party" while offering moral support to ultra-conservative clerics under fire from the left. This talent had enabled him to recruit alternatively from conservatives, the middle class and from disillusioned socialists and workers. The latter, as Haider wrote in his book:

...understand our message concerning social security reform, privilege and foreign immigration. Austrian workers were never really leftist in outlook. The former chancellor and Socialist Bruno Kreisky recognised this. Just before his death he invited me for a long talk. 'They don't understand the ordinary people anymore,' he said with obvious reference to his successor Franz Vranitzky. Kreisky himself had always struck a chord with the 'little man' and the break with this tradition in his own movement greatly grieved him. His remarks stuck in my memory as I had the impression he was trying to communicate something to me for my own political course. My movement has become a broadly-based citizens' movement for conscientious, decent and open-minded people. The betrayed Socialist as well as the middle-class intellectual will both find a home with us. The bridge between the workers and intellectuals has influenced the national-liberal camp ever since the revolution of 1848.[2]

Haider, Austria's answer to Robin Hood, was to rob the "fat cats" to give to the poor. He used Kreisky's phraseology and appealed to all who wanted to go a part of the way with him on this road to renewal. Understandably many Socialists found this tactic galling and offensive, but some, such as Hans Janitschek, former general secretary of the Socialist International and Günther Rehak, an ex-secretary of Kreisky, were impressed with the Freedom Party. The SPÖ was a party of the new Establishment. It was too closely associated, as Haider was fond of saying, with the bankers and the "golf course set who make politics with a champagne glass." As such, the SPÖ was one of the most "conservative" parties in Austria seeking to preserve the status quo and its own power centres. The SPÖ had been in opposition for only four years out of the first fifty in the Second Republic. Even under Kreisky, the "sleaze" factor started to dog the SPÖ, revolving around an insider clique within the party network. Much was revealed in the books of Hans Pretterebner, a maverick journalist, who in 1994 was elected to parliament on the FPÖ ticket.[3]

As the Kreisky era came to a close, many in the SPÖ were fearful that the party had lost its way and was out of touch with ordinary people. The left wing, numerically weak and not influential, bemoaned the "Tuscany set" in the party whose understanding of "red" was limited to wine tastings in Viennese vinotheks. The fire of the pre-war Austro-Marxism was burnt-out and as even the former head of the Industrialists' Association, Herbert Krejci, once put it, "the SPÖ has become a cold party without a heart."

Haider's appeal to the "man on the street" was necessary to get his party out of the electoral ghetto in which it found itself under Steger. This also admirably fitted Haider's style of trendy, telegenic politics. That Haider should be the logical successor to Bruno Kreisky was not at first sight obvious. In Britain it could be argued that Labour's Tony Blair did not look the natural continuation of Lady Thatcher although he came across that way: both stood out against the old class-ridden Establishment and wanted "security through change." As Blair said in his Hayman speech in Australia in 1995, "Thatcher was a radical not a Tory." He shared similar values with Thatcher such as law and order and an emphasis on the family. Like Thatcher, he fought against vested interests and privilege. He also supported radical constitutional reform which had widespread implications for the monarchy, the House of Lords and the Union with Scotland. Many who voted for Thatcher, Blair concluded, were not conservatives but "anti-Establishment" who saw "part of the Left as well as the Right running that Establishment." These voters could be targeted by "new" Labour.

The radical anti-Establishment card was played too by Haider. In Austria, however, the Establishment for the FPÖ was essentially dominated by '68-vintage pseudo-left Viennese intellectuals who were securely entrenched in influential positions in the Second Republic. For voters who wanted to protest against "those up there" the Freedom Party with its Third Republic was to be an attractive alternative.

Haider was often criticised for running a one-man show on authoritarian lines. The danger of the cult of personality is not unique in modern politics. In Britain, Blair came under fire from his own party for "a Stalinist leadership style." According to Richard Burden, a Labour member of parliament,

> Power is increasingly centralised on the leader's office, with immense pressure on everyone else to fall into line in the interests of unity and not jeopardising electoral chances. I am worried about the prospect of a party continually concerned with avoiding the spread of negative images of itself, desperate to be elected...with an inner sanctum holding a virtual monopoly on defining what mainstream opinion consists of. I thought that kind of approach to political leadership went out of fashion when the Berlin Wall came down.

Burden claimed that Labour under Blair is "drifting towards a US style party, a ruthlessly effective electoral machine as the vehicle for those who want to go into politics." Ideology has been relegated to glossy packaging and marketing. Blair had a Clintonised hairstyle and professional speeches. The "American Way of Politics" had crept into Europe.

In Austria there were some in the SPÖ who were also disturbed by what they considered to be a lack of freedom of thought. Franz Ramskogler, leader of the party's Young Generation, considered that whatever did not "fit into the mainstream of the banker" was unwelcome.[4] He complained about the lack of discussion in the party executive: "The chancellor gives a report – there are some comments and that's it." This criticism came from the ranks of a party whose leadership consistently attacked Haider for authoritarianism.

Haider believed along with Gingrich that civilisation was based on a spiritual and moral dimension with an emphasis on personal responsibility as well as individual rights. Both believed that this civilisation was endangered by suspect elites who wanted a "culture of irresponsibility" incompatible with freedom.[5] The Freedom Party like the Tory right in Britain had common ground with Gingrich when he wrote: "We must replace the welfare state with an opportunity society. Too many people are bound in bureaucracies and anti-human regulations by which families are destroyed and the work ethic undermined." The transference of power from the centre to the periphery and individual citizens was another common theme. In Britain it was Michael Portillo, a Thatcherite standard-bearer who in 1995 became defence minister, who put forward a similar blueprint to renew Britain. "Our highest goal is to minimise the size of the State, so that the individual can increasingly take responsibility for his own actions," he wrote in 1993. "The State," he continued, "has slipped into an attitude of studied amorality.... Our system tends to treat alike the unfortunate and the feckless, the thrifty and the profligate. Consequently it undermines the provident and demoralises the industrious."[6] Like Haider he believed that most people were "decent, hard working and law abiding. They value personal independence. They want to keep a fair proportion of what they earn.... Of course they want to help the needy. They do not, however, want to help those who are better off than they are, nor those who make no effort to help themselves."

Some cut-backs on public spending advocated in Britain and America were more stringent than proposed by the FPÖ. In this sense Haider was softer than some of his contemporaries. In Austria the claimant culture had stronger roots but gradually it became clear that the days of the bloated welfare state were over. After the election of 1995 the new government started to tackle some of the abuses in the welfare state and cut back on items such as free in-patient rehabilitation treatment and family allowances. Prescription charges were increased and employees had to pay for a form before they could visit doctors or dentists on the national health lists. Although there was no fundamental restructuring of social security or the health system as demanded by the *Freiheitlichen*, the ethos of some of its reform programme was in effect endorsed by the SPÖ/ÖVP.

Haider belongs to the same generation as Blair and Portillo. He shares their flamboyance and talent for provocative sound-bites and aggressive one-liners. There were many similarities in their philosophy but Haider lacked respectability. Instead he was compared more often with Schönhuber in Germany and Le Pen in France. Many Austrians could see the pressing need for changes in the country and were critical of the party book system, the rigidity of the political system and reluctance to innovate. They would begrudgingly concede that much of what Haider said was right. They feared, however, a Haider who had grown up in the "brown" (Nazi) milieu in a country where democracy had relatively new and potentially shaky roots. The blast of fresh air was in principle welcome but for them it came from the wrong quarter in a country whose contemporary history was bespattered with fascism and the experience of Nazi occupation. The people, it was believed, were vulnerable to manipulation by a character like Haider who floated the suspect idea of a presidial republic with an injection of plebiscites. This was not the "liberalisation" of Austria they had in mind.

Historical parallels were often drawn between the F movement and the rise of the Nazis in the 1920s and inevitably with Adolf Hitler. What the American historian Max Riedlsperger called "the lingering shadow of Nazism" continued to follow the third camp well into the 1990s. In Austria to be on the "right" meant either to be associated with German nationalism or Catholic fundamentalism. An Anglo-American "conservative" was, in the Second Republic, politically homeless. Similarly "out" was the brash challenger to the status quo and accepted orthodoxy. Despite his punchiness, Haider resisted the political savagery of some of Britain's politicians. Austrian journalists, too, generally lacked the impudence and nerve of some of their Anglo-American counterparts when dealing with government politicians. The consensus republic, with its low conflict culture, interpreted criticism as a declaration of war.

TRANS-ATLANTIC TRIP

With his increasing successes, Haider discovered America. He made several trips to the continent to study at first hand the immigration issue in California and to promote contacts and the English version of his book, *The Freedom I Mean*. During his summer holidays in 1996 Haider pursued a course in economics ("Global Reform and the Privatisation of Public Enterprises") at his own expense at Harvard University. It was important for the party to promote a more respectable image abroad and to cultivate contacts with the right people in Washington and New York. This had been relatively neglected by the party which had concentrated on building up its position within the country.

On one of these visits in 1995 Haider was introduced as the leader of the Austrian Liberals to the infuriation of Heide Schmidt who demanded an explanation in parliament. The leader of the Freedom movement also made waves back home by describing his native Austria as an "underdeveloped democracy" with shades of the Kremlin and the Vatican. Some considered that this would damage the tourist trade and the economy. Others deplored Haider's speech which they maintained dented Austria's image in the USA painstakingly built up after the Waldheim affair. Haider had justified his remarks by saying that in Austria there were limitations on freedom of expression. He attacked the TV broadcasting monopoly and repeated many of his criticisms which he had previously made in campaigns and in his book. To do this outside of Austria, however, seemed for many tantamount to treason. It was just not the "done thing." Haider also took the opportunity whilst on tour to drop in to the Museum of Tolerance, the Simon Wiesenthal Centre in Los Angeles. The F leader was interested to check out his portrait which had allegedly been placed along with politicians such as Idi Amin and Le Pen. For Haider this did more damage to the Austrian image abroad than his lectures (see interview section of this volume – interviews with Haider and Wiesenthal).

One of the criteria put forward for the "right extremist" classification was "criticism of democracy." Haider was considered by his opponents as a borderline democrat because of his attacks on the Austrian democratic system. When he spoke of a democratic deficit, many accused him of wanting to do away with the system altogether. The deficiencies in Austria and limitations on freedom had been highlighted by Schmidt's Liberals during the 1994 election. They put up posters showing mouths gagged with barbed wire and sticky tape. They also attacked the ban on protest demonstrations in Vienna during the visit of the Chinese prime minister. Even the broadcasting monopoly had been condemned as incompatible with human rights by the European Court of Justice for Human Rights in Strasbourg. Political commentators and journalists often remarked that some television broadcasts bore a resemblance to former communist manipulated news in Eastern Europe. An "interview" with the then chancellor Vranitzky in 1994 on the government's budget proposals was considered to be more like a party political broadcast. The weekly encounter with the press in the federal chancellory was similarly described by a leading journalist as a kind of propaganda exercise for the government. On one occasion when a reporter insisted on repeating an embarrassing question the chancellor walked off: the cut and thrust of rough political dialogue was not customary in the Second Republic.

Whatever the strengths or weaknesses of the Austrian Second Republic's democracy, a cool detached discussion was more complicated after any statement by Jörg Haider. During the 1994 election campaign Haider sparked off a furor when he expressed the view that "representative democracy is out-of-date." In an interview with the daily *Der Standard* (31 August 1994), he made the by now predictable attack on the "party state." The future, according to Haider, belonged less to representative democracy and parties in the traditional sense than to electoral movements and citizens' rights movements.[7] Haider was accused by Chancellor Vranitzky of wanting to abolish representative democracy altogether. Haider's commitment to democracy and democratic institutions was questioned by the then general secretary of the ÖVP, Ingrid Korosec, who suspected Haider wanted to set up a presidential-*Führer* system. Haider replied that he did not propose abolishing parliament but wanted to see a democracy where people and not parties were important. Direct democracy had to be strengthened to achieve this goal so that citizens would have a say on a wider range of issues such as taxation and immigration. The current state of democracy Haider described as wretched and "contaminated by the parties." Haider's critique of the Second Republic went to the quick of everything that had been built up in Austria after 1945. He mercilessly exposed its raw spots including archaic structures, neo-corporatism and monarchical relics. The trouble was, as many of even his most ardent opponents would confess in private, that he had a point. What they failed to do was to draw up a plausible alternative to remedy those aspects of the Second Republic which, after 50 years, were in need of an overhaul.

After the 1994 election the Freedom movement went on the offensive on the issue of democracy, confident it had the support of 22 percent of the electorate and was determined to no longer be marginalised. Haider was at some pains to defend his democratic credentials especially after a spate of letter bombs – suspected to have been masterminded by the extreme right – once again put a question mark over the sympathies of some in his movement.

In a basic policy statement in March 1995 entitled "We Democrats," Haider gave unequivocal support to democracy. "Who profits from terror?" he asked. "Not us! We democrats triumph through the ballot box in peace and freedom!" He reserved the right, however, to expose flaws in the system when freedom was endangered. He cautioned his opponents who liked to see themselves as liberal-minded against intolerance when it came to *his* movement. He appealed for more fair play and less hatred towards his person and supporters. For Liberals and many Social Democrats, however, it was Haider who had created the climate in which hatred and political extremism could flourish. They blamed him for a primitive manipulation of people's fears and of deliberately playing up emotional issues such as immigration. Each side accused the other with equal vehemence of destabilising democracy in Austria.

Notes

1. For a full discussion of right extremism, see H. Wohnout, "Rechtsextremismus, Rechtspopulismus und ihre Rückwirkungen auf das österreichische politische System," in Khol-Ofner-Stirnemann, eds., *Österreichisches Jahrbuch für Politik '93* (Vienna: Verlag für Geschichte und Politik, 1994), pp. 381-401.

2. Related in *The Freedom I Mean* (New York: Swan, 1995), pp. 67-68. Haider's claim to be the true heir of Kreisky poured salt on deep wounds in the SPÖ. Many Socialists felt that the ideals of Kreisky had been abandoned by his technocratic successors.

The account of this meeting is disputed by Social Democratic contemporary historians. Kreisky was an ambivalent figure for many in the SPÖ – he had led the party to electoral victories and stamped an "era" on Austrian politics. Towards the end of his life, however, he had made his displeasure clear with his successors such as ex-banker Vranitzky. On television in November 1996, Haider insisted that such a meeting had taken place between himself and the former chancellor and how they had reconciled their former differences. Haider recalled how Kreisky had foreseen that there would one day be three middle-sized parties of which the SPÖ would just be in the lead. He told Haider how important it was in this situation to keep bridges between these three for the sake of Austria (see also interview with Haider in the interview section of this volume).

3. H. Pretterebner, *Der Fall Lucona* (Munich: Knaur, 1989).

4. *Die Presse*, 7 June 1995.

5. Newt Gingrich, *To Renew America* (New York: Harper Collins, 1995).

6. Michael Portillo, *Clear Blue Water* (London: CWF, 1994).

7. The federal president Dr. Thomas Klestil took up this theme in a speech in September 1996 pointing out that the idea of representative democracy no longer was convincing. Democratic institutions had become alien for many Austrians who felt that politicians were increasingly remote from the needs of ordinary citizens. This led to an increasing number of non-voters and protest voters. Klestil suggested a "republic of citizens" was necessary to redress this imbalance. Democracy for the president had to be more than "party democracy." He supported therefore more direct democracy and directly-elected mayors. (Speech given at the ORF conference on "The Sovereign Citizen?" 5 September 1996.) Klestil's theory was immediately refuted by leading SPÖ and ÖVP parliamentarians.

5

Programmatic Development

The Freedom Party had a conventional political programme which had been adopted at a conference in Salzburg in 1985 under the leadership of the liberal Norbert Steger. This remained official policy for the *Freiheitlichen* well into the 1990s. The concept of the Third Republic was informally grafted onto this in addition to other ad hoc basic policy declarations. A new basic programme was drawn up in 1997.

The following extracts illustrate the programmatic development of the FPÖ:

Platform of the Austrian Freedom Party, 1985

Freedom

- Freedom is the value most precious to us. We seek a life in freedom, based on self-determination, independence and personal responsibility. Our politics aspire to a way of life with the greatest possible self-determination for each person and for all peoples.
- Our liberal mission is to place the freedom of the individual and his or her dignity in the top rank of the social order. Our national mission is to place the freedom of peoples and their self-esteem at the top of the world order. Our ecological task is to protect the free development of nature from subordination to the technological and economic purposes of mankind.
- We want to protect freedom from physical, intellectual and economic oppression. We want at the same time to prevent freedom from slipping into a "free-for-all" or into anarchy.
- Respect for the liberties of all men requires that freedom be safeguarded within a framework of order. Political structures must serve to develop freedom. Since freedom also presupposes responsibility, liberal systems strive for a balance of rights and duties.
- The society of the free can only endure if vital common tasks are addressed. Our politics affirm the responsibility of free people to undertake essential duties in the service of the people, one's country and the state.

HUMAN DIGNITY

- Human dignity demands our absolute respect. We want a tolerant society, which ensures each person the expression of his or her individuality.
- Each human being is unique as an individual, yet fundamentally endowed with equal rights.
- We oppose enforced benevolence and totalitarian transformation of people to fit a preset mould. We support an open pluralist society where differing points of view and life-styles exist together. Variety enriches human life whereas uniformity impoverishes it. The recognition of such diversity does not justify differences in the estimation of the dignity of the individual.
- The freedom of an individual is limited by the freedom of others. For everybody is part of society, which supports them and in return imposes obligations on them.
- We consider man and woman to be of equal worth, of equal standing in society and therefore, equal in their degree of responsibility. We strive for a partnership between the sexes in all aspects of life.
- We regard the family as the most important element in the community. All families deserve the protection of society. We respect the demands of the family and its autonomy. This autonomy is conditioned by the respect for the personal rights of individual family members.

PEOPLE AND HOMELAND

- We regard the democratic republic of Austria as our mother country, in which the values of national tradition and love of one's country are to be highly valued. We favour a national policy which permits ethnic groups to preserve their basic rights and allows them to express their own identities.
- Family and a people have organically grown and demand respect in politics. People, as a natural community, are bound together by heritage and history and have a common language and culture. We consider the existence of ethnic communities to be necessary, especially at a time of super-national and super-regional integration. The co-existence of ethnic communities irrespective of national boundaries must be safeguarded in the spirit of mutual respect and tolerance.

- We reject national arrogance. We condemn every abuse of nationalist fervour for totalitarian or imperialist ends. The tragic consequences of the abuse of nationalist ideas, particularly in this century, are horrifying examples from which we must learn for the future.
- The overwhelming majority of Austrians belong to the German ethnic and cultural community. This fact remains even though it has in many ways been suppressed as the result of a disastrous chapter of German history in Austria. We want an Austria, embedded within the German ethnic and cultural region, to autonomously shape its future.
- We appreciate the existence of ethnic minorities in Austria as a valuable cultural enrichment of our common homeland. We are for generous protection of minorities on the basis of valid ethnic rights.

EUROPE

- The future of Europe lies in a close community of all its countries and peoples. In spite of all the difficulties of unification the goal remains a unified and strong Europe, to which there is no reasonable alternative. The last step to unification is, in our view, the creation of a European confederation. In this confederation each people ought to be able, on the basis of self-determination, to preserve its identity.
- We seek the unification of Europe to safeguard political and economic survival as well as to ensure peace in the world.
- More community in Europe must never mean less freedom for its citizens. To protect ethnic minorities and to address minority issues, we favour a European law for ethnic groups based on the right to self-determination and right to a homeland.

CULTURE

- From a national viewpoint, our cultural policy affirms the cultural self-identification and self-expression of ethnic groups and all peoples. Recognising that Austria forms part of the German cultural region, the cultivation of German culture and language is of special importance to us. We consider it our duty to provide an Austrian contribution to the development of German culture. This notwithstanding, we acknowledge from a national perspective the cultural identity of other ethnic minorities and want to help them to secure and develop their cultural heritage.

SOCIAL COMMUNITY

- We demand social justice, not only in terms of equality before the law but also of equality of opportunity. Differences which result from achievement and hard work must not be leveled down. Forced egalitarianism is the enemy of freedom.
- We want more freedom and less government but not the renunciation by the state of its social responsibility to the poor and the weak.
- In the society we stand for, freedom is associated with consideration for others and communal spirit. Social welfare institutions must ensure that anyone in need is guaranteed at least a minimum level of care. On the other hand, the abuse of welfare institutions must be prevented. The guiding principle of our social policy is primarily the encouragement of self-help. We therefore reject a social policy which creates permanent dependence.

ACHIEVEMENT

- Hard work and achievement should remain the driving force behind social development. We regard the will to achieve as an expression of the desire by the individual to actively control his or her destiny in all areas, be they intellectual or physical, cultural or social, technical or economic.
- The desire for achievement must not be suffocated by undue pressure. Those who do not put so much stress on this goal must not be discriminated against. Achievement for the common good deserves special recognition.
- The desire to achieve is often accompanied by a willingness to take risks, accept sacrifices and break new ground. Boldness of this sort, be it on a large or small scale, is also in the interests of the development of society, and should therefore be encouraged. For this reason all forms of recognition and the earnings structure should be oriented primarily towards the achievement principle. Achievement must be worthwhile. Persons should not be deprived of the fruits of their individual labour. We therefore reject ideologies based on the egalitarian distribution of wealth.

- Competition is a permanent feature of our pluralist society. We approve of competition but demand rules to prevent immoral behaviour and the abuse of power. The formation of open elites on the basis of actual significant achievements safeguards the viability of great human societies. Elites must continuously prove themselves. We reject the idea of privilege.

PROPERTY AND MARKET ECONOMY

- The recognition of private property is a basic condition for any free society. We want as much property as possible in private hands. We reject the nationalisation of property. Economic tasks assumed by nationalised companies or co-operatives or local authorities should be continuously checked for their reprivatisation potential.
- We reject exploitation. Property rights may not be exercised without consideration for society. Above all ownership of land, means of production and capital in general are subject to social and ecological considerations.
- We favour a liberal market economy with as much freedom of action and autonomy as possible for the individual entrepreneur within a social and ecological framework.
- Any type of centrally controlled state economy is incompatible with a free social order.
- The main task of our free market economic policy is the maintenance of fair competition. A freedom-oriented economic policy aims at establishing a harmonious balance between all market forces and particularly between capital and labour.

GOVERNMENT AND JUSTICE

- We believe in a liberal, democratic republic, in the principle of a state under the rule of law, in a multi-party system and in the free competition of all political forces. For us democracy means the power of the people exercised fundamentally through majority decisions of its representatives selected in free general elections on the principle of proportional representation. In addition we believe in self-government by the people and therefore demand the development and refinement of the instruments of direct democracy.

- Individuals can only be expected to accept ethical values in the community if they have confidence that those responsible in government particularly political officeholders, also act in accordance with these principles. A sense of duty and freedom from corruption are essential if citizens are to have confidence in their government.

ENVIRONMENT

- The total exploitation of nature by man in the course of scientific, technological and economic development endangers the ecosystem world-wide. By rendering himself as the master of nature, man has turned into its destroyer. He is in danger of bringing about his own extinction through the gradual destruction of the necessities of life.
- Since this is a task of global dimensions, a new environmental awareness must be fostered in people of all nations.
- A recycling economy must be created in which the reuse and the reprocessing of materials in short supply will be taken as a matter of course.
- All aspirations for a global environmental policy will ultimately fail if the growth of world population cannot be halted. We favour a stabilisation of population figures through humane birth control provisions.
- Citizens should be involved from the beginning in decisions that substantially affect the environment. Environmental protection is not without cost. It might also require sacrifices and affect traditional life-styles.

An interim upgrade to this basic programme came with the Freedom Party's "Theses for a Political Renewal of Austria," passed in December 1993, under the leadership of Haider, and adopted as the electoral platform of the FPÖ in 1994.

The Theses conspicuously omitted references to the German cultural community and its *Volk* and stressed Austrian patriotism. The party was out to net conservative-minded voters normally inclined to vote for the People's Party. The Theses adopted a hard line, militant position on law and order, immigration and defence.

THESES FOR THE POLITICAL RENEWAL OF AUSTRIA
(Extracts)

1. Change is urgently needed. The rampant growth of privilege and sinecure stands in contradiction to a liberal philosophy. We stand for the creation of an open society with the courage to allow for diversity based on a sense of responsibility.
2. The Freedom Party is the only force for change in society in Austria. The aim is to liberate the citizen from political parties. The spheres of domination by the old parties on jobs and apartments, which serve to keep people dependent, should be done away with. In this sense the Freedom Party regards itself as a large citizen rights movement. Austria should no longer be a republic of the Reds and the Blacks but the home of Austrians. This is a battle against a system of oppression.
3. There can be no freedom without order. However freedom means responsibility. Responsibility of the individual is the best protection from domination. The greater the sense of responsibility, the less necessity there is for regulatory intervention.
4. There is no freedom without law and there can be no law without justice.
5. The recognition of Austria as our homeland is the basis of our political action. We want to preserve and maintain the natural environment and cultural heritage inherited from our forefathers in order to hand them on to our descendants. Love of one's native country corresponds to respect for the home countries of other peoples. This positive form of patriotism is the answer to excessive nationalism and chauvinism on the one hand and the utopia of a multi-cultural society on the other. It thoroughly promotes the idea of being European.
6. We stand for a liberal democratic republic on the basis of human and citizens' rights. These rights cannot be abrogated, even by democratic means. Democracy guarantees the highest degree of liberty and justice as all law must originate from the people. However democracy must not lead to the oppression of a minority by the majority or to marginalisation. Tolerance is vital.

7. We stand for the direct election of mayors and provincial governors; for strong direct democratic instruments following the proven Swiss pattern. We want members of parliament who respect the will of the citizen, the decentralisation of power and federalism.
8. All necessary means of force must be used in the fight against terrorism and the drug trade.
9. For an Austria of Austrians. Uncontrolled immigration is pushing the limits. The protection of cultural identity and social peace in Austria requires a stop to immigration. Compulsory proof of identity, rigorous controls, increased border security and systematic deportation of illegal immigrants are necessary to ensure the effectiveness of residence and immigration laws.
10. We stand for the continued existence of a sovereign Austria in a pan-European confederation of states which guarantees all citizens security and which accords the people freedom of self-determination and the right to their own home. A functioning security system be it new or an adapted NATO, should have absolute priority.
11. We stand for a substantial build up in defence and a professional army. A high degree of professionalism and the latest, modern equipment can be achieved by a voluntary army consisting of professional soldiers and voluntary militia.
12. We stand for the preservation and improvement of the environment by an ecological tax reform – charging for the consumption of natural resources while at the same time reducing taxes on human labour.
13. We stand for a judicial system which puts the victim before the criminal and are determined to fight the rise of crime with sufficient personnel and with effective sentences.
14. We want a competitive social market economy based on private property. We want the reduction of taxes and the privatisation of all state-owned companies, administrative reform and a balanced budget.
15. We want to cut down the party political bias of the social partners and trade unions. We stand for social policies which do not create dependency or encourage the abuse of institutions.

16. We fight for the maintenance of the farming community through agricultural and forestry policies which guarantee the protection of agricultural structures and the maintenance of family farms.
17. We stand for health policies which guarantee a performance and cost-oriented system and more private insurance.
18. We stand for a performance oriented education system without party influence and socio-political experiments. We favour aptitude tests and free choice of schools.
19. For us the family is the core of a society in which young people should have equal opportunities and the elderly should not be pushed out. We stand for reduced taxation of families.
20. We want affordable housing by the creation of competitive conditions with no interference from political parties.
21. Nationalised industries are to be privatised under conditions which allow companies to remain competitive and avoid sell-out to foreign investors.
22. We stand for a culture which reflects the free development of citizen and society. It is the state's responsibility to preserve our cultural heritage; contemporary art has to take account of public criticism and not political influence. State support should be limited to preserving plurality in education and the arts.
23. Monopolies, state or party control over the media are hostile to democracy. Journalism means responsibility.
24. We stand for the preservation of natural ethnic groups and the protection of their cultural identity. However such protection is not to be extended to new immigrants. Austria is not an immigration country.

This platform was adopted after the rift with Heide Schmidt and her decision to break with Haider and start a Liberal Forum in February 1993. She and former colleagues who had been in the FPÖ abandoned the party depressed at the lack of scope they saw for their liberalism. The FPÖ replied that her move was heavily encouraged by the SPÖ in parliament and prompted by personal career motives. Soon after this the FPÖ left the Liberal International (LI). Relations between the two had been strained after Haider became leader of the Freedom Party. Representatives of the LI had visited Austria to follow the FPÖ's campaigns, disturbed by what they saw as illiberalism especially on immigration. The break with the LI was a logical development since for the FPÖ the organisation was remote-controlled by a clique of leftists.

The old style ultra-German nationalists in the FPÖ were also unhappy with the way the party was moving. They saw an insidious dilution of ideology and a relegation of the importance of the German culture. Many, like Kriemhild Trattnig from a rural peasant background in Carinthia, had little time for the "yuppie boys" who had the confidence of Haider. Both representatives from the liberal and the German national wing of the FPÖ saw an increasing tendency to ditch principle in favour of vote-catching.

"AUSTRIA ÜBER ALLES"

In 1997 a fresh effort was made to hammer out a traditional party programme to replace the 1985 Salzburg programme. Work on this was in progress in 1995 but had been interrupted by the unexpected general election. The task was resumed under the aegis of a leading parliamentarian, Ewald Stadler, and an elite inner circle of party intellectuals.[1] After several brainstorming sessions in the Tyrol, a draft programme emerged which it was hoped would be adopted at a special conference in June 1997. This was too fast for some members, especially in the Vienna party, who wanted more time to reflect and debate sensitive issues. The conference was therefore delayed until late October 1997 to meet these demands.

In the preamble to the draft programme entitled "Austria first," the party declared itself to be the only credible guardian of Austrian patriotism: "The Freedom movement puts Austria, the country and its people, above everything else especially party political interests." The movement felt obliged to defend the interests and identity of Austria which it believed were being increasingly undermined. The party was described as a political force for reform which stood for fairness, equality of opportunity and rights of the citizens against rigid corporatism, institutionalised class warfare and the tyranny of a political caste system.

The lay-out of the programme consisted of brief statements followed by a longer commentary. The statements were to be the foundation blocks of the programme which would only be changed by a party conference. The commentary was to be more flexible to allow for a quick response to developing issues and problems and could be changed by the party leadership. After some debate, this idea was dropped in favour of a uniform format as in the past.

The first statement defined freedom and self determination as the most precious values of the Freedom movement. It categorically rejected every type of oppression and arbitrary state action. This concept of freedom was to be accorded to individuals and the family, ethnic groups, peoples as well as the community; it was not to degenerate into a cult of egoism. A sense of responsibility and duty was necessary for freedom to flourish. The importance of private property was stressed for freedom to have any real meaning for people.

This was followed by a section stating the inviolability of human dignity. Everyone was to be allowed to develop their own capabilities in an open, pluralistic society. Physical and intellectual discrimination as well as political marginalisation were condemned outright. Diversity was to be defended against ideologically-motivated efforts by state institutions to set up dependence and uniformity.

A special chapter in the draft programme was dedicated to the idea of Austria as the Fatherland. Austria was not a mere accident of fate but its peoples were bound together by a cultural heritage. This was the basis of a new Austrian patriotism centred around a democratic, federal republic which was complemented in turn by a new term "constitutional patriotism." This was a major innovation compared with the old 1985 programme and it neatly skated around the problem of pinning Austria to the banner of the German *Volk* and cultural community. In a separate section on culture and art the importance of language for culture was stressed. It was therefore implicit that the cultural community for Austrians was determined by their mother tongue which was, in most cases, German.

After some internal discussions, the word Fatherland was abandoned and substituted with the less problematic word "Homeland." The draft programme further defined this right to a homeland as consisting of ethnic, territorial and cultural components. This included the democratic republic of Austria, its provinces and long-established ethnic groups (Germans, Croats, Romanies, Slovaks, Slovenes, Czechs and Hungarians). Their traditions and heritage were to be developed and protected. In this connection the draft programme cited the Basic Law of 21 December 1867 on the General Rights of Nationals in the Kingdoms and *Länder* represented in the *Reichsrat*, article 19: "All the ethnic entities of the empire enjoy equal rights and each ethnic entity has an inviolable right to the preservation and fostering of its nationality and language." According to the FPÖ every Austrian was to be free to decide their own identity and ethnic group. It was logical that as most Austrians spoke German this would be the largest group but the programme stressed there was to be discrimination by the state, whatever the choice. This section concluded with a statement that Austria was no country of immigration on account of its topographical situation but that it would remain a country for asylum-seekers fleeing from religious, racial or political persecution.

"ONWARD CHRISTIAN SOLDIERS"

A controversial part of the draft programme dealt with a new, positive relationship of the party to religion and the Church. It was partly because of confusion on this, that a decision was made to postpone the special conference on the programme. Many, especially in Vienna, were sentimental about the traditional anti-clericism of the party and were suspicious that it was becoming too pro-Catholic. Chief co-ordinator Stadler, a staunch catholic himself, was at pains to refute this explaining that while Christianity as a whole was an important spiritual foundation for Europe, other confessions and religions including Judaism had also historically made an important contribution. The spiritual values of the West were to be defended through co-operation between the different churches in Europe. This implied a kind of militant Christianity in which the Freedom movement would play its part. The aim was to counter the dangers posed by Islamic Fundamentalism and the growth of pseudo-religious sects which were insidiously undermining the values so precious to Europeans. Unlike the Schmidt Liberals the FPÖ wanted to keep religious education in the syllabus in schools. The draft programme made it quite clear that anti-clericism was completely obsolete. The concept of a militant Christianity was dropped after opposition from sections of the Vienna party. It was replaced by a reference to a "Christianity which would defend its values." The phrase "Islamic Fundamentalism" was changed to "religious Fundamentalism."

Following this was the idea that Europe was more than a geographical entity but had been shaped by Christian, western values and a common cultural heritage. This amounted to more than the sum of the supra-national institutions of the European Union. The future of Europe lay in diversity based on the collaboration of essentially independent states. A European currency was seen as the final stage of a long process of economic integration and rejected for the time being as too premature.

Neutrality was deemed obsolete with the entry of Austria to the European Union and membership of NATO was supported as well as the build up of a professional army.

A FREE REPUBLIC

The term "Third Republic" had been open to misunderstanding, so many of the proposals for constitutional reform, as envisaged in the "Third Republic," were subsumed under a section dealing with a "free republic." This was to lead to more accountability and transparency and less party domination and bureaucracy. The free republic spelt out the party's commitment in more detail to parliamentarism and democracy. It conveyed the impression that the FPÖ was a fully constitutional party seeking constructive reforms instead of a revolutionary overthrow of the system. The programme supported "law and order" measures but bluntly rejected the reintroduction of the death penalty.

The new republic was to be free politically and economically fair. While the ÖVP had once agonised over terms such as eco-social market economy which no one could really grasp or pronounce, the FPÖ went for a "fair market economy." This was to be free, socially just and ecologically friendly. It was to reward the thrifty and diligent and further private enterprise independent of state and party controls. It could appeal to trade unionists and libertarian free marketers alike and received little attention from the press and party functionaries.

It was questionable whether a conventional programme was really appropriate for the new F movement which could arguably get by with a shorter action programme. The main programme of the party anyway was "Jörg Haider."

NOTES

1. I am grateful to Ewald Stadler for a preview of the draft programme and to him and Lothar Höbelt for sparing the time to discuss some of its main points. It is noteworthy that Haider entrusted the job of drafting a new party programme to someone so prominent in the FPÖ hierarchy. In the subsequent intra-party controversies, Haider cleverly played the role of the "honest broker." The intense discussion of Stadler's document showed that the FPÖ was capable of engaging in intellectual debate on points of principle and that compromises could be made. The final adoption of the party programme came in October 1997, too late to be included in this volume.

6

HAIDER'S CREDO

Trying to track Haider's own position was like shooting a moving target. The F leader variously appealed to the workers, the middle class, farmers, German nationalists, Austrian patriots, anticlerics and staunch Catholics. He made a virtue out of what some called inconsistency and claimed simply to be on the ball in an ever-changing world. His political gyrations were impressive. No one could so convincingly claim that what at first sight looked like U-turns were really a straight road to the same goal. The FPÖ was one of the first parties in Austria to support membership of the European Community (EC). Under Haider the F adopted a passionately Eurosceptic, but arguably pro-European position. It said "yes" to the EC but rejected "Maastricht," i.e., full political union with common social, foreign and currency policies

In addition to major policy statements, speeches and interviews the Haider credo is summed up in his book *Die Freiheit, die ich Meine*, published in 1993 and which appeared in 1995 in English as, *The Freedom I Mean*. The title was taken from a song composed during the Napoleonic wars at a time when students from Germany and Austria were fighting for freedom. This book was a controversial contribution to the discussion on the "Third Republic" which Haider described as a "new era free from party patronage and despotism." It included a discourse on the author's concept of freedom and the limitations on this which in his opinion existed in Austria.

Throughout his writings it is clear that Haider was disturbed by what he saw as the moral collapse of society with the family as the core unit based on Christian principles. An over-regulated state, a media monopoly and excessive individualism were responsible for this moral disintegration. According to Haider, "once puritanical abstinence and the Protestant ethic were effective correctives to excessive consumption. Work had no negative connotations and was seen as the fulfillment of professional duty to make a contribution to the common good. This has given way to a plastic credit card society in which no-one thinks any more about what they could do for others."

Haider called for an end to the "cradle to grave society" where everything was laid on by the "nanny state" for those with the party book. This stifled, in his opinion, individual initiative and freedom. He supported however the social state which would provide for the weak and infirm who were genuinely in need of help. "Hangers-on" such as party functionaries who played the system for their own ends were to be given short shrift. Any reform had to deal with the oversized bureaucracy which had to be pruned and made more efficient. His philosophy was reminiscent of Gingrich on this point when he wrote,

> Hard work must pay off. But the collectivist welfare state rewards all, irrespective of performance, in the same measure. The 'achievers' are in effect penalised and the bone idle rewarded. A society where people study longer, work less, earn more, have more leisure time and retire early is totally unrealistic. We must have the courage to remunerate those willing to work and contribute. This will make it unattractive for passive fellow-travellers to abuse the system.

Haider managed to appeal both to those in society who were losing out and to those active elements with initiative and drive who wanted to "get on."

Freedom, for Haider, had been substantially curtailed in Austria by the cultural hegemony of the "bankrupt '68 generation." This was seen as "the ruling political class" whose "goals are to destabilise, dissolve, devalue and marginalise." Haider's judgment on this "clique" is harsh. These people, he writes, "cultivate their insatiable ideological needs and conjure up artificial enemies. With their modern Inquisition they produce an intellectual climate of crippling conformism. Intolerance of those with different opinions is typical...in this scheme of things, the so-called populist is an unwelcome guest."

This power set-up had firmly shut the door on Haider and his party. He was grieved but unrepentant. His withering critique of the left and Austrian socialists shows a penetrating insight – "they are 'conservative' in so far as they stand for their own naked power claims and the preservation of their own influence. They are no longer bothered about a better world or a just society but only with keeping their own jobs and positions...their revolutionary spirit dried out in the sun-baked vineyards of Tuscany." For the

Freedom movement the "pin-stripe banker" Vranitzky had become out of touch with his own people. Haider claimed that despite a long period in office the SPÖ had not tackled wage differentials or radically redistributed wealth. He maintained that labour organisations were often more generous with foreign workers than with Austrian working women. "I have met," he writes, "many on my visits to factories who earn less in a month than our politicians get for travelling expenses...why should they care? Most of them have so much, they don't know what to spend their money on while the small farmers, their wives, pensioners, women workers and all those who can't find work over 50 are the Cinderellas of our Socialist society."

Haider was in the vanguard against "political correctness," another American import to European culture which many thought had gone too far. The silent majority found in Haider a spokesman who stood up against the progressive wave and ivory tower immigration policies and feminism. The right had begun its own emancipation. It called for a halt to immigration and a condemnation of Stalinism as well as National Socialism. Political decency was no longer to be a monopoly of the so-called left intellectuals. A favourite target of Haider in this crusade was the "multi-cultural dreamland." He believed a society must rest on a shared value system or risk drifting into chaos and anarchy. Ideologically perverted multi-culturalism for Haider was unrealistic since "it is not the immigrants who integrate into the society and culture they find themselves in, instead they expect from the natives that they should accept their customs. Peaceful integration on these terms is not likely." As an example Haider cited the outcry when Islamic parents demanded that the crucifix be removed from schoolrooms in Austria because it offended the religious feelings of their children.

Feminist fantasies were also lampooned by Haider who believed many women were forced to go out to work for financial reasons. "Women should have the choice between family and career, free from ideological trappings.... The local creche or Kindergarten is no substitute for a mother...instead kids get fast food, stress and pre-packaged education via television." Those left feminists Haider had in mind immediately charged him with wanting to send women back to the chains of the kitchen. For Haider the family was a secure anchor for children and its collapse had led to a generation hooked on drugs and attracted to violence. Tax policies were to be revamped to allow women the freedom to look after their children.

Haider was convinced his programme for political renewal was in tune with the times and that the future belonged to him and his Freedom movement. He saw that elsewhere in Europe the tide was moving in a similar direction, "the parties and governments in most countries in Europe are isolated from their people, stranded on a barren desert island. The general malaise with the old parties knows no borders." Haider saw his mission in Austria to preserve freedom, regenerate a flagging democracy and preserve common values based on western civilization. This was his vision and own particular brand of freedom.

VIENNA DECLARATION

A milestone in the evolution of the thinking of the FPÖ leader came with the "Vienna Declaration," a paper delivered in April 1992. This coincided with the momentous changes in Europe following German reunification and the fall of the Berlin Wall. For Haider the end of the post-war era in Europe heralded the end of post-war Austria. The country had now to redefine its position in Europe but was rooted to the past by the dogmas of the old parties. It was up to the Freedom Party, Haider believed, as a reforming force to tell people the truth even if it was not popular. It was time to end the deadening climate of conformity weighted against dissent which the two main parties had created: "too long the old parties have ruled this country without serious competition.... Too often Austrians have had to sit back and watch how the institutions of the republic were degraded into self-service shops for the parties." Haider railed against "compulsory membership in compulsory professional societies" and the ubiquitous and infamous party book system.

Haider then replied to charges that he wanted an authoritarian *Führer* state. "We have a helpless government, weak established parties and an all-powerful bureaucracy. So if a politician appears in this country who shows that he can actually take decisions and put an end to fruitless talking in circles with no solutions, he is promoted straight-away to the rank of an authoritarian leader."

For him consensus-oriented democracy could not just be coincident with the needs of the old parties. He ruled out force as a political means and categorically rejected every form of totalitarian, chauvinistic, and National Socialist thinking.

In the Vienna Declaration Haider made plain his loyalties were first and foremost to Austria. This was to counter the frequent allegations that he and his party were prone to hyper-Germanness. Haider spoke of the Austrian contribution to German history and insisted on a specific Austrian identity. To put it simply, he said, "I am first of all an Austrian with body and soul; everything else takes second place." He was clearly irritated by his opponents who claimed to have a monopoly when it came to standing up for the sovereignty and independence of Austria.

Haider had already sensed that the "sectarian confrontation between national and liberal" was a sterile debate. The time had come to end such ideological fundamentalism and dogma and build a broadly-based freedom movement. He was to return to this theme and tried gradually to wean his followers from an outmoded pan-German complex.

Haider concluded the Vienna statement with an appeal to all those who could go some way with him in building a new republic free from party domination:

> Whoever goes with me stands for an Austrian Freedom Party without brown stains but also without fear of historical truth.
>
> Whoever goes with me stands for an Austrian Freedom Party with credible distance from the time of National Socialism but with respect for the older generation, which found the way to democracy after their own bitter experiences.
>
> Whoever goes with me stands for an Austrian Freedom Party which is committed to the German ethnic and cultural community albeit with the proviso that the commitment to Austria is unmistakable concerning its identity, the inviolability of its borders and on the condition that its sovereign existence is not put in question.
>
> Whoever goes with me stands for an Austrian Freedom Party which does not represent materialistic but rather cultural ethical values which lay the basis for the protection of minorities and entail a clear distance from racism and anti-Semitism.
>
> Whoever goes with me stands for an Austrian Freedom Party which wants a free state based on the rule of law and which creates an open society and thus rejects closed social systems such as national fundamentalism or liberal individualism devoid of cohesion.
>
> We are, so I believe, on the right path. Long live the republic! Long live our fatherland Austria!

Extracts from this speech were printed in the daily *Die Presse*, 8 April 1992. Despite this declaration, Haider was constantly accused of not making a clear statement on the National Socialist period. Many found such statements lacking in credibility in view of Haider's brown "lapsus linguae," but they nevertheless can go on record.

THE AUSTRIA DECLARATION

Haider's conversion to the Austria idea was maturing by the time of the election in October 1994. As part of his platform he delivered a major speech entitled "The Austria Declaration." It contained the familiar onslaught on the media monopoly and "thought control" which for Haider was akin to Orwell's *1984*. The coalition in Austria between "the Reds and the Blacks" had degenerated into one-party rule by a "unity party" under Socialist control. "Socialism wields power in the government, in the ORF, in the constitutional court, in the judiciary, in the National Bank, in the Trade Union Federation and in the Post Office Savings Bank, in the Building Cooperatives, in culture, in the Chambers of Labour and in social insurance." The People's Party was seen as the pawn of Socialism, obediently acquiescing in this accumulation of power.

Haider accused the entire government of complicity in squandering tax payers' money for its own ends. Many people, Haider deduced, were fed up with this state of affairs but resigned to it continuing. They felt helpless against a powerful party cartel. This is where the Freedom Party stepped in as "the gladiators, the minority in a political arena, who fight against the Lions in power." This colourful language was designed to arouse sympathy for the underdog. It evoked the imagery of David against Goliath and proved electorally successful. Haider at other times appeared like a knight in shining armour on the side of the losers in an affluent Austria who worked hard to make ends meet. He stood up for the "ordinary" decent people such as working mothers, hard-working small businessmen and commuters who had to travel to find work. He contrasted their tough life with the easy number of Socialist bosses with their smart villas, generous severance pay and chauffeur-driven cars subsidised by the "man on the street."

Haider knew the "ruling political class" had been weakened by his rise but warned ominously, "the Empire fights back without mercy." He bemoaned the "psycho-terror" of the red/black Establishment which could not tolerate those who thought differently. Haider against the rest of the world and the Austrian "evil empire" could be sure to pull in support. The election slogans of the party were similarly effective: "Simply honest, simply Jörg" and "They're against him because he's for you."

THE "GERMAN QUESTION"

Haider's understanding of history moved from the assumption that the workers had never really been "left" in Austria. In the 1994 election, many workers, disillusioned with the Socialists, went over to Haider and his Freedom Party. They responded to themes like "law and order" and sought security in a troubled world. The FPÖ had expanded beyond its traditional constituency and Haider liked to think of it as the real "Austrian Party" of the people. The task afterwards was to keep new converts in the fold and to offer them something interesting in return for their continued support. In addition, Haider wanted to give his movement a respectable image which would tempt more voters to go over to him and which would have a positive resonance abroad. The emphasis on the "Austrian Way"[1] fitted into these ambitions. The popular initiative organised by the FPÖ on the question of immigration ran under the slogan "Austria first," and the party's "no" position on the European Union referendum was defended from an Austrian patriotic standpoint. Slowly Haider had been edging his followers to a more pronounced pro-Austria position and away from Teutonic roots. Haider's own contacts with Germany anyway had become less rosy. He disagreed with Chancellor Kohl's vision of a future Europe and was booed by hostile crowds during a tour of Germany.

In the summer of 1995 Haider felt the time was ripe to risk a rupture with the hard-core German nationalists in his party. He revealed in a magazine article that *Deutschtümelei* had no place in his movement.[2] This concept was associated with an excessive fondness for the "German cultural community" as laid down in the 1985 programme but also implied a glorification of all things German. This kind of hyper-German nationalism was regarded as ideological ballast from the past which could be embarrassing or damaging. Haider's comments opened the door for a programmatic debate on the "German question." Predictably, these remarks provoked an outcry from the few militant fundamentalists on the German nationalist fringe. They believed that Haider was either no longer in command or at best purely obsessed with getting into power whatever the price. Only around five percent of the electorate could really be mobilised by the German factor; the rest found it irrelevant or too academic to be interesting. For Haider the point was whether the party was equipped to move forward to the next century without the dead weight of a bygone age. The Vienna organisation gave muffled support to their chief on this point, while the strongest critics were to be found in Carinthia amongst old German nationals who had long since ceased to play an active role in politics. These people were particularly bitter as they had stood behind Haider in his crucial leadership bid in 1986. For them Haider was the prodigy of the German nationalist wing against what they saw as the liberal lunacy of the Steger clique. As they saw it, their ideals were now being coldly ditched in a power game where the stakes were much higher.

In Britain, Tony Blair was faced with a similar problem involving the unreconstructed comrades on the left who equated socialism with public ownership. This principle was enshrined in Clause IV of the Labour Party's constitution which dated from 1918. It had long ceased to have any practical relevance but had a symbolic importance cherished by many in the rank and file. Blair proposed making a clear break with Clause IV and opened up a debate in the party which many considered unnecessary and divisive. He, too, was accused of sacrificing principle in his drive for the premiership. Some argued that it was better to sink the hard-liners before getting into power rather than have them drift on to scuttle the party in soffice.

Haider's motives often aroused suspicion and puzzled even his followers. Although successful in elections, polls showed that few wanted to really see him as chancellor. Many admired him as a tough-nosed politician but he was left with the reputation of a shady second-hand car dealer who could not always be trusted. He needed to stay in the limelight and thrived on media attention. This necessitated frequent and controversial sound-bites. Endless press bulletins and new announcements, however, gave the impression of inconsistency and lack of careful planning.

Haider's *Deutschtümelei* observations were accompanied by a sudden disinterest in the new title "F" for the movement. The "F" title had been widely adopted by the press and subject to much misunderstanding and ridicule. After barely eight months it was regarded as a marketing flop and casually dropped by Haider who replied to an interviewer, "I don't like it much either."

The general response to Haider as *Homo Austriacus* was sceptical. For the leader of the People's Party in parliament, Andreas Khol, Haider was just scraping the surface: "As long as he seeks a Third Republic, as long as he holds pre-fascist thoughts like having a strong man instead of the federal chancellor and president, he stands outside the pale of the constitution." Khol underlined the shifty perception of Haider by many: "Whoever trusts in you, is building on sand." Haider still remained, officially at least, a man with whom in Austria "nobody would do business" even if he tried to cultivate a more moderate image.

His opponents and the media still insisted on a pigeon-hole for the leader and his *Freiheitlichen*. Once a party which had embraced a German national and liberal core element, the F (alias FPÖ) now seemed bereft of both. What survived was the Haider credo, a cocktail of ideas which both the media and Haider agreed could be termed "right radical populism" (ORF interview, 20 August 1995). With the increasing decomposition of *Lager* loyalty, this label had the merit of being suitably vague and eminently flexible. The "F" movement was a heterogeneous incorporation of several groups. These included the hard-core pan-Germans, others with decidedly "brown" sympathies (not to be bundled together with the first group), left-over liberals who could not bring themselves to embrace Schmidt's band, the discontented wanting to register a protest, achievers and non-achievers, and lost neo-conservatives. This motley collection no longer amounted to a coherent "third *Lager*" or encampment which was confined to an electoral dead end. It was supported by over a million voters and could not afford to become the captive of any one group. The whole enterprise was held together by the ingenious versatility of Jörg Haider and a constant ebb and flow of diverse ideas. The problem was that, as with any largish group, core principles were in danger of being diluted to protect the "catch-all" element of a modern "people's party."

"ADOLF HAIDER?"

Not all were convinced by the portrayal of Haider as the knight in shining armour who would slay the red/black dragons of a corrupt republic. For many he was a political "bovver boy," the skinhead from Carinthia, or a brown wolf in sheep's clothing. Haider remained an enigma.

Those who distrusted Haider and believed the gloss of democracy went only skin-deep could cite a number of infamous Haider "clangers." One of the most widely reported of these abroad was his remarks on the Nazis' employment policy in June 1991. This was not a special speech dedicated to the subject but was rather an interlocution in a debate in the provincial diet of Carinthia. Haider was criticising the labour market administration and according to the protocol went on record as saying the following:

> "...and he pays with his hard-earned money higher deductions so that then a few can have a good time in the hammock of the social state. This is no system which we can defend." [Applause from the FPÖ; interjection from the SPÖ floor leader Hausenblas: "What you're demanding has already been done, but in the Third Reich!"] Reply from Haider: "No, that didn't happen in the Third Reich because in the Third Reich they had a sound employment policy...." After about a quarter of an hour an uproar broke out. Haider makes it clear that "I unequivocally made the point that this remark was not made with the meaning understood by you. If it reassures you then I take back the remark with regret."

The damage, however, had been done and Haider was forced to step down as provincial governor in Carinthia a few weeks later. The Socialists seized on the comment with glee. They were facing a party conference which lacked a theme and the "Nazis' employment policy" came as a gift. Ranks closed around Vranitzky who battled against Haider with the same conviction as he had stood out against Waldheim. "Stop Haider" became a unifying slogan which covered up the paucity of ideas in the Socialist Party. The coalition partner ÖVP also condemned Haider's slip of the tongue. It was generally interpreted as playing down the evils of National Socialism, the Holocaust, concentration camps, forced labour and economic policies which led to war and destruction. A motion of "no confidence" in

Haider was moved by the SPÖ in Carinthia and supported by the ÖVP. Haider's two year reign as governor in the province was over.

Provincial governors in Austria have powers to appoint staff in the administration, schools and hospitals. It can be a congenial and rewarding job and governors can often become popular patriarchal figures. To lose such a jewel clearly wounded Haider and he vowed revenge at the polls at the first opportunity. His followers saw his departure as the work of the "governmental execution commando" to eliminate a successful rival. On the eve of his departure around 10,000 showed up in Klagenfurt, the capital of Carinthia, for a solidarity rally with Haider. The case re-opened a debate on the Third Reich and the part played in the Holocaust by Austrians from different political parties. It did little to contribute in a constructive way to the subject of the original debate on unemployment and abuses in the welfare state.

For many Haider was finished. He became a political leper, shunned and vilified and deemed unfit to hold high office. He was condemned by the leaders of all parties represented in parliament, the federal president, bishops and the media. It was not the first time Haider had made comments of this kind. He had once referred to the Austrian nation as an "ideological miscarriage"[3] and denied that the FPÖ was a successor to the Nazi party "since if it was it would have an absolute majority." It was not the last time either that he was carelessly to drop remarks which provided his enemies with useful ammunition.

"PENAL CAMPS"

Another remark which fueled charges of a problematic relationship with the Third Reich came in the so-called "penal camp" speech. During a parliamentary debate on terrorism in 1995 Haider talked of "the penal camps of National Socialism." The speech was devoted to a condemnation of violence and intolerance following bomb attacks in Burgenland in which four people died. Haider spoke at around midday and was followed by speakers from the Greens and by the Liberal Heide Schmidt. They had either not listened attentively to the speech or had not found anything immediately amiss. By four o'clock in the afternoon, however, the speech had become a burning issue and the reference to "penal camps" was seized on by a Green deputy to show that Haider was playing down the atrocities of

Nazism by neglecting to use the term "concentration camps." "Penal," it was argued, implied that those who suffered in these camps were guilty and fulfilling some kind of sentence as a result. Haider claimed that for him the terms were interchangeable and that of course by "penal camps" he had meant "concentration camps." In the same debate later in the evening, a Green member of parliament (Gabriela Moser) also referred to penal camps in a similar connection: "In this situation, ladies and gentlemen, we should not let ourselves be catapulted back to the year 1945. In 1945 there were people in penal camps who never deserved punishment, who never committed a crime but who on the contrary had been working for a basic democratic consensus like Rosa Jochmann whom we have already mentioned today." No one got excited about the reference to "penal camps" on this occasion, although Jochmann was arrested by the Nazis and taken to the concentration camp Ravensbruck. In the stenographic protocol of parliament, however, the speech was amended to include after "penal camp" – "gemeint sind KZs" (i.e., concentration camps). The FPÖ complained of a crass manipulation of the text whilst the Green concerned insisted that this was only a proper amplification of what she had meant to say. Corrections are allowed subsequently to parliamentary speeches but these are not supposed to substantively alter the content.

In April 1995, the federal president, on the occasion of the republic's fiftieth birthday, made a speech in parliament in which he referred to the "dungeons" (*Kerkern*) of National Socialism. There was no comment from Haider's opponents to these statements and no charges of playing down the evils of Nazism.

The heart of Haider's speech dealt with some embarrassing anomalies for the governing parties. He drew attention to the uncomfortable fact that the plight of the victims, members of the Romany ethnic group, had long been neglected by both the ÖVP and SPÖ. When the deaths of these four people hit the headlines, promises of improvement were made. The funerals were attended by what many cynically saw as the big-wigs from Vienna out to get their picture in the paper. Six months later as Green member of parliament pointed out, in August 1995, nothing tangible had emerged from the pious words of the government.

In the same plenary session of parliament during which Haider talked of "penal camps," another incident occurred which illustrated the tense and emotional atmosphere in the chamber. A member from

Haider's party referred to an attempted attack on his leader during an election meeting in Linz. Someone had thrown a container at Haider full of an explosive mixture of weed salt. The deputy floor leader for the Socialists, Ilse Mertle, interrupted with an aside – "that would be a possibility" – and even repeated the remark when asked by the FPÖ to clarify her strange statement. The Freedom members in parliament were outraged by this apparent nonchalant endorsement of attempted violence which could have endangered life. The Greens and the Liberals seemed to be unconcerned by the remark. The ÖVP accused the Freedom members of artificially exaggerating the problem out of all proportion.

By now it was difficult to judge *bona fide* statements by either side. Haider, whether intentionally or not, polarised opinion into two distinct groups – either for or against him. Increasingly the Freedom members were confronted with an "unholy alliance" or the "gang of four" of the other parties represented in parliament (SPÖ, ÖVP plus Greens and Liberals). Often Haider succeeded in achieving the impossible – that of creating a common platform between the otherwise diverse parties arrayed against him. It was a scenario which allowed Haider to score points, and present voters with a clear alternative. By contrast the other parties merged into an anti-Haider coalition with little constructive to offer.

THE CENTRE FRINGE

Haider's "slip-ups" continually cast doubt over his credentials as a good democrat and as a politician opposed to National Socialist totalitarian ideology. The roots of his party and the connections of some members with the right extremist fringe meant that he was constantly on the defensive on this vexed issue. His remarks were often deliberately misconstrued by the opposition to damage Haider's image with the public – in politics a familiar technique. Some surmised that such dubious comments were not accidental lapses but deliberately staged by Haider to pander to the tiny pan-German national element in the party. Other theories supposed that deep in Haider's psyche lay a tiny "Adolf" trying like Houdini to break free. Critics remarked that typical for such Haider gaffes was the accompanying tactic of half-apologising or qualifying the offending remark a day or so later in an effort to retain the good will of the middle-of-the-road voter. In this way Haider walked the path of the centre fringe, always careful to return to the straight and narrow at the right time.

On the subject of the Third Reich, as on many other topics, Haider was no wild adherent of political correctness. When the Socialist interior minister Einem suggested offering political asylum to women who had been victims of rape in the war-torn Balkans, Haider rubbished the idea. When the ÖVP Vice-Chancellor Schüssel, on behalf of the government, offered to double contributions from the general public which would go to help those in former Yugoslavia, Haider sardonically pointed out that this was tax-payers' money and if the cabinet wanted to show generosity it ought to come from ministers' salaries instead.

It was difficult for the other parties to squarely dump the burden of Austria's past at Haider's doorstep. Not all former Nazis had ended up with the Third *Lager*. Some had even moved up the ranks of the SPÖ and became cabinet ministers in Kreisky's first government. During the Waldheim affair, members of the ÖVP were accused of appealing to anti-Semitic prejudice.[4]

The coded use of common Nazi terminology was interpreted by some as a deliberate signal to those on the extreme right fringe. Deliberate and frequent use of such diction, it was said, was employed to awaken old prejudices. Some compared Haider's tactics with those of Hitler and pointed out that he surrounded himself with young men who owed their careers to the "chief." Without Haider these people were nothing. Also many found it suspicious that Haider would turn up late to party rallies allegedly to heighten the atmosphere and stimulate the emotions of the audience. Heide Schmidt was renowned for being late but this, in contrast, was put down to an inability to be punctual. When Haider showed up late it was just as likely to be due to disorganisation, chaotic management or simply because he was stuck in a traffic jam or had to follow a tight schedule rather than the result of a studied imitation of Hitler and the Nuremberg rallies. Sometimes Haider even turned up early to events and had to go off for a bite to eat or a beer until the crowd came.[5]

Haider was at some pains to clean up his party and erase any possible charge that it could be lumbered with the politically damaging right extremist title. He broke with the German national fixation, stressed all things Austrian and extended an invitation to Jews to join him. He disassociated himself from extreme right publications and asked his adherents to avoid meeting extremists on the right. "Haider has always attempted to appeal to various constituencies simultaneously, and psephological exigencies have required that his strategic opening to the political right be tempered by the vociferous protestation of his party's commitment to constitutional legality."[6] Haider wanted to project himself as the model democrat but for many he still remained a dangerous bogeyman.

If Haider could be accused of using inflammatory language, the same could also be leveled at his opponents. Chancellor Vranitzky called Haider in 1987 a "party political rat-catcher" and the then vice-chancellor Erhard Busek referred to "a political dwarf: here quarantine is appropriate so that the rotten bacillus dies" (1992). The Green Peter Pilz considered that "the FPÖ belongs in the beer tents and not in parliament." The Socialist Josef Cap accused Haider of creating a climate of hate and mistrust and condemned his confidants as "hooligans of Austrian domestic politics." Haider's number two in the "F" hierarchy in parliament, Ewald Stadler, renowned for his acerbic speeches, was compared to a vicious Doberman dog. During one of Haider's speeches in parliament a Green made an obscene gesture to the Freedom leader who replied, "That isn't the fine English manner."

A consistent charge of Cap was that Haider was a political arsonist with fascistic rhetoric who wanted to overthrow democracy and establish a *Führer* system. In return Haider called Vranitzky an "Austro-fascist" who could not tolerate those who thought differently.

The attempt to box the FPÖ into an extreme brown corner proved a failure. Another strategy tried by the coalition of trying to overtake Haider on the right, such as on the issue of immigration, proved equally futile as the elections of 1994 showed. It was difficult to see where the two main parties really stood. The ÖVP sporadically opted for the "politics of the centre" while the SPÖ toyed with the idea of liberalism. Both clung together in coalition for fear of letting Haider loose in government. This had the effect of cementing the post-war rigidity in the party system, the opposite of what Haider sought to bring about. Haider could claim success in getting the big parties to react to his policies whilst the coalition, by keeping Haider out of business, could feel it upheld Austria's famous (or notorious) stability.

NOTES

1. "The Austrian Way" was a highly potent election slogan from the Kreisky era. The message was slowly being adopted and adapted by Haider who was trying to portray the F/FPÖ as the Austria party par excellence. He projected a movement which stood out against multi-culturalism and left Austria to the Austrians. It sought a country which also looked to the proud traditions and customs of the provinces in addition to the urbane smoothness of the capital. It built on the culture of the native fatherland and would fight off creeping EU irredentism.
2. Interview in *Wirtschaft und Politik*, August, nr. 34, 1995. For the problem of Austria's relationship with Germany, see G. Holzer, *Verfreundete Nachbarn. Österreich — Deutschland. Ein Verhältnis* (Vienna: K&S, 1995).
3. In a television interview in the summer of 1995, Haider put on record that he would no longer use this term.
4. Also on this theme, see A. Pelinka, *Windstille* (Vienna: Medusa, 1985) and A. Pelinka, "Die Großparteien und der Rechtsextremismus," in *Handbuch des Österreichischen Rechtsextremismus* (Vienna: DÖW, 1993), pp. 464-73. Also, A. Pelinka, *Zur Österreichischen Identität* (Vienna: Ueberreuter, 1990), pp. 51-58 and 84-88. Pelinka describes Leopold Kunschak, honorary chairman of the ÖVP and president of the Austrian parliament, 1945, as a "radical anti-Semite" whilst Karl Renner, Socialist president of the Second Republic, showed "more understanding for 'small Nazis' than for displaced Jews" (p. 61).
5. See Hans-Henning Scharsach, *Haider's Kampf* (Vienna: Orac, 1992). Also based on discussion with Anneliese Rohrer of *Die Presse* and discussions with FPÖ party officials.
6. *Anti-Semitism, Word Report*, "Austria," published by the Institute of Jewish Affairs (London, 1995), pp. 80-81.

"PENAL CAMPS"
Speech delivered by Dr. Haider in the National Council

Ladies and Gentlemen,

People have been shocked by the recent terrorist attacks which claimed the lives of four Austrian citizens and seriously injured another.

These events have made parliament and all its groups subdued. Subdued because the terror which seems to be seizing our country is being expressed in an increasingly disgraceful manner.

The first letter bombs tried to spark off distress in this country and we can say those attacked got off lightly. It was very different later however with the attack on the Mayor of Vienna who sustained serious injuries as a result of this terror.

I consider it wrong when via the press service and through newspaper statements somebody or other tries to draw hasty conclusions or settle political accounts; for in the final analysis we are today in the midst of a discussion in which Austrians as a whole will be accused. This is gradually going in the direction of seeing Austrians, as it were, as an intolerant people and not open-minded if such things can happen. And when I read today that the Austrian police are supposed to be the intellectual and political allies of rightist extremism in Austria, then this is going a bit too far.

Ladies and gentlemen: these insults concerning our country in no way solve the problems which are confronting us. Our task must be in this situation to find a common bond of all democratic forces in the recognition that this type of debate represents a new quality in Austria.

This type of terror has never happened before and every upright democrat, whatever ideological and philosophical reservations he may have against one or the other must say: we must draw the line here together on violence, and so effectively, that people in this country do not have any reason to be afraid and can feel safe.

The line must be drawn together against all those who sympathise with violence, for violence, ladies and gentlemen, has no ideological ribbons; it is neither left nor right – it is contemptuous of human beings and leads to awful murder.

Our task today can only be this – to say: here the gauntlet has been thrown down to a democratic society based on the rule of law. We take up this gauntlet of violence and make it unmistakably clear that we will not indulge in abuse, or in mutual recriminations, but that we will act decisively and give a clear signal that terror in this country has no chance and that we will in no way succumb to violence – nor will we yield until these crimes and acts of violence have been, without exception, cleared up. [Applause from the FPÖ.]

Therefore, Mr. Minister, it is somehow unsatisfactory when you said yesterday – I've had enough. You have raised many hopes. On 14 December 1993 you talked of a hot lead. On 24 February 1994 you said: at last a big tip-off concerning them and described how the perpetrator behind the bombs was resident in West Germany. You announced on 9 October 1994 the first real tangible lead. You stated on 10 October: there are strong clues concerning the makers of the bomb. You announced on 11 October 1994 that there was a "red hot lead" in *Die Presse* and likewise you talked of a hot clue in this affair on television. And then – nothing. And this is what makes people feel unsafe: that simply statements are being issued which obviously leave you groping around in the dark without anything really transpiring. Just try once more to engage the entire force of the executive order to clear up these terrible attacks. Why can only a special squad do this which is anyway hopelessly overstretched? Why don't you mobilise in all federal provinces all forces and why don't you try in this way to get the necessary success in clearing up the matter? Today we don't want to take you to task and say: you are to blame or you have done something wrong. But it is absolutely vital before putting out bulletins to have a think first and see if you have any shred of material which you could make known. And if you know about the scene, then you must act, and it should not happen as it has since 1993 until now that hot leads are constantly announced but no result transpires.

And I should also like to say something else: it is very nice when we stand up for ethnic groups. We Austrians have long-standing ethnic groups.

But we must not forget that a co-existence of the majority and minority in this country was always seen as an ideal target for terrorists in order to bring about a destabilisation here.

I should like to remind you that in the 1970s in Carinthia we had 12 bomb attacks also resulting in people being hurt. It transpired after the twelfth bomb that this series of attacks had been carried out by the Yugoslav secret service to destabilise the situation in Carinthia. With the twelfth bomb the terrorist blew himself up as an observant museum curator ordered the museum to be cleared just in time and the culprit panicked.

I quote here an article from "Delo" which appeared in Laibach on 25 November: "What has happened in recent months in Carinthia is an old story seen with reference to the bombs appearing only in a new chapter. Remember how Belgrade for years sent Slovenian agents to blow up Slovenian partisan monuments with dynamite in different parts of Carinthia. What for? To sow hatred between Slovenes and Austrians! I should like," continued the writer, "after all to go even further and mark out a triangle: Serbs–Nazis–Italian fascists, since it is well known for how long and in what close contact the Italian information service has with every Serbian exponent."

Ladies and gentlemen: we should not be too shortsighted on this question. We should recognise that apparently a network is on the move here which works on a European level for the destabilisation of democratic states based on the rule of law. There's more behind this than we may assume at first sight.

And I'd like to add something else: this treatment of a minority is also a key which could possibly reduce the fun of terrorists. We had until the end of the 1980s a latent conflict situation in Carinthia between the ethnic group and the majority of the population. I should like to take some credit during my time as provincial governor for bringing about a lasting peace in these relationships. There are no more conflicts. I should like to present a very concrete question to the politicians of the ÖVP and the SPÖ. Burgenland is a federal province which is led by a Social Democratic governor. Oberwart is a community which had an ÖVP mayor. Why is there such a discrepancy in the treatment of the Sinti and Romany minority in practice from that which is formulated here? Why are the Sinti and Romanies even in 1995 settled outside the Oberwart boundary? Why are they not integrated, if what you say here is meant to be taken seriously? Why is there no real dialogue between the minorities, the Croats and the majority of the population in Burgenland? Why have those measures which have been so valuable in Carinthia not been carried out in Burgenland? What use here are statements in parliament when those who have power in politics in practice are not prepared to do something concrete for an improvement in these relationships? [Applause from the FPÖ.] They should think about this.

For the non-integration of an ethnic minority, which already once before 50 years ago was almost exterminated in the penal camps of National Socialism, and then was resettled and marginalised, is linked with the fact that the will which is manifested here is not translated into practical politics. Things like the "sea of light" are no help at all. The nice statements which have been made today are useless! The assumption is that no-one can attempt to split the majority from the minority or to set them against one another. It would be much more important to answer the question why right up until today there is no road link between the village and the Sinti and the Romanies. Why is it a goods route? Is that not just an expression of the way we deal with each other? Do we not support something here which in the practical politics of those who have made such statements, cannot be made up for nor carried out? Do we not profess tolerance when in reality the minority suffer legal disadvantages in their basic interests in life?

For, ladies and gentlemen, so long as there is this discrepancy and the possibility of a conflict situation between the majority and minority and we allow it, there is a great danger that a destabilisation is possible by terrorist activities.

We Freedomites want to take part not just in explanations but also in strengthening this democracy in the clear understanding that the fortification of a democracy shows itself through a capacity to tolerate the opinions of others without making a criminal of them.

I am addressing this in particular to those who like to criminalise the Freedomites because they can't counter them with political success.

The type of moral intimidation which some here in parliament throw in the direction of my party makes it sometimes impossible to freely make points of view. Or as Martin Walser said this type of terror which is practised by some and above all leftist groups makes free speech and thought a dangerous and risky exercise. Just consider also whether it is appropriate following these events not just to understand others but also to respect them. That's true not just for minorities but also for dealings with one another. [Applause from the FPÖ.]

••••

The FPÖ issued the following press statement 9 February 1995: "The leader of the Freedomites today made it unequivocally clear that in his references concerning the extermination of Romanies under National Socialism he clearly meant the heinous National Socialist concentration camps."

7

"AUSTRIA FIRST"

"Austria First" was the title of a popular initiative (petition drive) introduced by the FPÖ on the subject of immigration.[1] It was also the campaign slogan of Thomas Klestil when he successfully ran for the presidency in 1992. The FPÖ initiative aroused charges of xenophobia and racism. Immigration became an explosive issue after the fall of the Iron Curtain particularly in and around Vienna, which was the first stop-off point for those moving westwards. Haider was accused of deliberately appealing to people's base instincts and exploiting fear of an influx of foreigners. The hard-line position of the FPÖ proved an electoral winner as shown in the Vienna municipal elections in November 1991. In some districts in the capital, teachers were confronted with classes where 80 percent of the pupils were immigrants. Many feared the growth of slums, ghettos and an increase in crime. The FPÖ quoted official statistics which showed a high crime rate amongst foreigners especially those from former Yugoslavia and Turkey. The alarming increase in drug dealing was also linked by the party to foreigners from Turkey and Nigeria.

The decision was taken by the FPÖ late in 1992 to go ahead with the controversial popular initiative, scheduled for the beginning of 1993. It was one reason given for Heide Schmidt's move to break with the FPÖ and set up her Liberal Forum in February 1993. She maintained that Haider was aware incidents could accompany the immigration initiative but decided to proceed regardless. Haider replied that he reckoned with provocation from his opponents and that the FPÖ could not be made responsible for this. Schmidt waited until after the results of the initiative were made public before breaking with the party, ostensibly because she did not want be made a "scapegoat" for any failure.[2]

Many believed immigration was not a topic which could be put to the people because it was too emotionally charged. The FPÖ countered that the people had the right to voice an opinion on such a vital issue which affected them, their standard of life, jobs and accommodation. The initiative included demands for a stop to immigration unless certain guarantees could be given on those points which would safeguard the native population. Haider always maintained that his position on immigration was to avoid the violence which had taken place in Germany spreading to Austria. He was not, as he put it, "anti-foreigner but pro-Austrian." Neither did he question the tradition of Austria to receive refugees according to the Geneva Convention and argued, "Austria is also a second home for many guest workers who with hard work and honesty have built up a living for themselves and their families. These people too will be endangered by uncontrolled immigration and could lose their livelihood" (in "Austria First," FPÖ documentation).

The FPÖ opposed the concept of uncontrolled immigration. The statement "Austria is no country of immigration" was intended to differentiate Austria from the classic countries which had received immigrants in the past such as America. However even in the "New World" the tide was beginning to turn. In the USA as also in other countries in Europe such as Britain and France, the feeling was growing that unregulated immigration could produce domestic conflict and was a burden on national resources. The Right in some US states such as California was tougher on the issue than Haider's FPÖ. There an initiative SOS (Save our State) sought to discourage the steady influx of illegal immigrants from Mexico, and central America. The Republican Right believed that citizenship was being exploited by pregnant women crossing the border who gave birth in American hospitals, guaranteeing their children access to US health care and education. The American Right wanted an end to automatic citizenship for children born to illegal immigrants on United States soil.

In 1995 Haider visited California to see the measures that were already being taken to deal with illegal immigration. The border in some places resembled an iron curtain with Mexico, with electronic sensors and strict patrols. An American citizens' movement called "Fair" wanted a new deal on immigration policy which went far beyond the FPÖ's initiative for Austria. They wanted to tighten up the border controls, refuse social security benefits for illegal immigrants and also envisaged a less privileged position for legal foreigners. The FPÖ "Austria First" campaign was regarded by many in California as too soft.[3]

Information circulated by the FPÖ during the petition drive contained statistics on immigration refuted by the Ministry of Interior. The FPÖ asserted that the percentage of foreigners living in Austria was around 12 percent of the total population whilst the ministry maintained the figure was only 6.6 percent. In Vienna according to the FPÖ 24 percent of the population were foreigners which contradicted the official figure of 12.8 per cent. According to estimations of the FPÖ (which had been derived from the socialist mayor Helmut Zilk) about 100,000 foreigners were illegally living in Vienna. The ministry replied that this figure was realistic but for the whole of Austria.

The number of signatures finally collected fell below the target set by the party (see Table 5). It had hoped for at least a million but had to be content with just over 400,000, a little more than half of the votes received by the FPÖ at the previous general election. Parliament duly dealt with the initiative but was not required to do any more. The initiative was deemed a "flop" by the media and regarded as a setback for the FPÖ. Party morale was at a low ebb when the Schmidt liberals abandoned ship soon afterwards forming a new parliamentary group.

The government heaved a sigh of relief, hoping that perhaps the Haider onslaught had been contained. To defuse the electoral appeal of the FPÖ on immigration, the coalition passed restrictive measures dealing even with foreigners who had for years been living in Austria. Bureaucratic and inhumane procedures caused hardship and suffering for many families. The Interior Ministry even threatened to rescind the residence permit of any foreigner who occupied less than 10 square metres of living space. A couple with a family therefore would have to live in an apartment which even many Austrians could not afford. Precise details of the apartment were required including the number of windows and type of plumbing. The rules also punished foreigners who could not submit the necessary documentation within four weeks of the old residence permit's expiration date including: an application form filled out in German, lease contract, health insurance certificate, work permit, police vetting sheet, marriage and birth certificates, aliens' registration certificate and proof of income. Many who had lived in Austria for decades suddenly ran the risk of expulsion. Foreign workers who fell sick or lost their jobs and were not entitled to dole money or health benefits were in a similar predicament because they could not prove they had a secure income. The authorities were unable to cope with the flood of applications which had to be made in person. Many applicants started queuing at the relevant offices early in the hours of the morning, only to be turned away midday when the offices closed and told to try their luck the next day. After an application had been filed a lengthy waiting time could be expected before a written decision was made. The information circulated to households drew attention to a special requirement for families: "A permit of residence may be granted to children under age and to foreign spouses of Austrians, or of foreigners who have stayed here for 2 years. If the quota is exhausted, application will have to be postponed until next year. Even Austrian born children to foreign parents have to have a permit." These measures were criticised by liberal-minded people and the Greens who suspected the government was trying to outflank Haider on the right. The government in effect took on board many of the points in the FPÖ Initiative. As the results of the 1994 election showed the attempt to knock out the Haider effect in this way was doomed to failure.

Haider and the FPÖ complained of a deliberate "hate campaign" against him and his supporters by the media and especially television. They spoke of a "psycho-terror" which effectively stopped people from signing the initiative. In contrast to a parliamentary election the instrument of popular initiative requires details of names and addresses to be given. Many who would support Haider in a secret ballot were reluctant to be seen openly signing what came to be called the "foreigner initiative," opposed by all other parties, the trade unions, the Industrialists' Association, the federal president, the church, artists and intellectuals, the World Wildlife Fund, Greenpeace and Global 2000. In January 1993 around a quarter of a million people took part in a "sea of light," a candle-lit demonstration in Vienna against xenophobia and racism organised by the "SOS for fellow humans." The demonstrators converged on the symbolic "Heroes' Square" where in 1938 Adolf Hitler spoke to an hysterical crowd after the *Anschluß*. This time the crowd was addressed by church dignitaries and children of foreigners who had settled in Austria. The federal theatre displayed a banner proudly listing the number of foreigners employed in the company and the main hospital complex also drew attention to the importance of foreigners with the slogan "medicine is international." Punks marched side by side with nuns and police who were allowed to take part in the demonstration. The sea of light was an unprecedented display of unity against an initiative that it was believed was being abused for party political purposes. The sea of light contributed to a lower turnout for the FPÖ's initiative than had been expected. To be seen voting for the initiative was not "politically correct" or considered the respectable, decent thing to do. Many voters were, according to analysts, alienated by the political style of the FPÖ which had attacked the federal president and the Church.

In his book Haider recalls the numerous letters he received from people in factories and homes for the elderly who feared the consequences if they had supported him. He described how posters in Vienna with information on the initiative were torn down or written over with the word "cancelled." Some mayors he alleged put the document to be signed in their own homes to deter support. Elsewhere the FPÖ claimed that buildings where signatures could be made were often locked. Some schools refused to act as voting offices on the grounds that lessons would be disrupted. Haider quoted the case of Burgl Czeitschner, head of one television department responsible for youth and the family, who issued written instructions during a meeting with her colleagues that "they should unobtrusively work against the initiative in all programmes." The allegation was confirmed by the ORF after Haider presented the protocol in a live television programme and Czeitschner was obliged to leave her post. The central secretary of the Social Democrats, Josef Cap, detected in Haider a "paranoiac style."

For Haider candles and pious words by intellectual lefties and idealistic "multi-cultis" were not enough to solve the desperate plight of many workers struggling for jobs and decent accommodation. Haider who had shaped the FPÖ into a trendy party of middle class yuppies with a liking for discos and fast cars, now went to the "ordinary man." He could increasingly be seen in working-class districts of Vienna where his message on immigration was immediately understood. For the SPÖ the immigration issue was an embarrassment and cost them the votes of the workers. Haider knew people were disturbed by classrooms and public transport where German was rarely heard. The Vienna coffee house intellectuals including Socialists and opinion leaders, so concerned about racial tolerance and integration, were known on the other hand to send their offspring to expensive private schools. The Socialists tried to reassure their normally disciplined and loyal voters. They promised them that municipal housing blocks would not be opened up to foreigners but reserved as always for native Austrians. Despite this, the workers were unsettled and felt they could no longer trust "their" party. "Liberals" on the other hand, won over to the SPÖ during the

Kreisky era, despaired of such gestures and gradually drifted away from the party.

Haider was convinced that ideological multi-culturalism was a dangerous experiment which threatened western values and customs. The spread of Islamic Fundamentalism particularly worried him as it clashed with western ideas on the rights of women and democracy. The need to protect native culture and traditions determined also his outlook on Austria's role in Europe. After the poor result for his immigration policies Haider announced that he would fight back and "not yield one millimetre." The setback seemed to strengthen his determination to carry on the fight. Haider's resolve to challenge an "authoritarian developing democracy" had won through before when he lost the governorship of Carinthia. It was to return after the disappointments in the 1995 election. These bruising experiences led to a more reflective and embittered mood. Haider complained of massive intimidation and moral blackmail by the fashionable elites of the Second Republic. They in turn saw in Haider a dangerous man with a persecution complex who had been beaten but not cowed.

Haider took up the gauntlet with relish and announced his next battle would be to mobilise support against the European Community. Afterwards would come the fight for the 1994 general election. In February 1993 Haider already prophesied that his goal was to get a million votes, or around 20 percent of the poll. In the aftermath of the results of the initiative, this seemed little more than a brave effort to rally the troops. This prophesy became a reality by 1994 and survived the setback of the premature election in winter 1995.

Not only did Haider survive to fight again after his initiative campaign, but the tenor of his thinking on immigration was adopted by some of his opponents. The deputy chairman of the SPÖ in Vienna, Councillor Hatzl, warned against sentimentality on the question of allowing children abroad to join their families living in Austria.[4] Whereas liberals believed it violated human rights to deny this, Hatzl considered it was inhumane to allow more immigrants to come to Austria where they would have to live in bad conditions.

Several abortive attempts to deal with the problem of immigration were made by the new coalition in 1996. A so-called "integration package" was drawn up which envisaged that children and dependents of guest workers could join their families in Austria. Otherwise priority was to be given to integrating foreigners already in Austria in preference to allowing more to enter the country. This worried the trade unions and the Chamber of Labour because of the pressure this would place on the labour market, schools and housing. The unions wanted to see more job creation schemes for younger people before further immigration. Industry too wanted to see a skilled labour force to make Austria competitive and feared the government's plans would hinder this development. Refugee organisations and the clergy were unhappy with another proposal which would make the fate of asylum seekers dependent on spot decisions by officials at the border. Haider threatened to launch another initiative in protest against what he saw as opening the floodgates to an intolerably high level of immigration. Re-uniting families for the FPÖ was to be achieved by the return of guest workers to their native countries and not by the shipping in of children and dependents to Austria. The FPÖ criticised the coalition package as irresponsible as the infrastructure was lacking to cope with family reunions. Warnings by the FPÖ of the growth of slums in Vienna were dismissed by the government parties as "alarmist." Yet in the districts most affected in the capital even socialist local politicians called for measures to spread the load by enabling guest workers to move to other parts of the city. They wanted immigrant children to attend schools in other districts with a lower percentage of foreigners. Chancellor Vranitzky rejected the idea of "bussing" foreign school kids around Vienna since it would only serve to spread the problem over a wider area instead of solving it. The socialist mayor of Vienna with a populist touch, Michael Häupl, opposed the idea of opening up municipal flats to foreigners because this would not be tolerated by the people.[5]

With an eye on the Vienna municipal elections scheduled for the autumn, the coalition withdrew the package for further review. Haider jubilantly called it "game, set and match" for his party's position on immigration. Not content with victory however he insisted on a special sitting of parliament to discuss the question in June 1996. Since the coalition seemed to be moving his way on restricting immigration, it was necessary for the FPÖ to go a step further to keep the upper hand. In the debate the party's speakers consistently made the point that there was a connection between the number of immigrants in the country and the number out of work, a point hotly contested by the other parties. As part of this one-upmanship therefore the FPÖ resurrected an idea floated before the 1995 election which allowed for repatriation of immigrant workers. This was to be open to unemployed immigrants who voluntarily wanted to go back to their native countries and who, in the opinion of the FPÖ, should be given some financial incentive to help them on their way.

A similar plan had been tentatively under discussion in Britain where it was mooted by a black Labour member of parliament, Bernie Grant, who represented the constituency of Tottenham in London. The idea was that immigrants from the Caribbean or Africa who wanted to return home should get some financial assistance from the government. At first sight this seemed a curious proposal to originate from the Labour benches but there were several categories of persons who apparently had shown interest in such a scheme. These included the long term unemployed who had given up the hope of a job in the UK or sick or handicapped people who wanted to go back home but lacked resources. Also there were young people who had families in Africa and without job prospects in Britain would welcome the opportunity to make a new start in Africa. Finally there were skilled and educated workers who still had relatives abroad and who wanted to return to make a contribution to building up the economy. This would involve a big cut in salary but government support would help and could be seen as a kind of development aid project. In the past these countries had suffered a brain drain and lost their qualified doctors and engineers etc. This scheme was a way of repairing this damage.[6] The FPÖ's proposals bore some similarity to the Grant scheme but predictably they were condemned as racist by the Austrian Greens, liberals and socialists.

A few months later under the new chancellor Klima, the government announced that financial assistance might be forthcoming to enable refugees (e.g., from Bosnia) to return to their home countries. The ÖVP Minister of Defence, Werner Fasslabend, deputising for Vice Chancellor Schüssel who was ill, suggested the scheme could be extended to include foreigners who were out of work. Suddenly the whole idea originally floated by the FPÖ looked humane and patriotic. Klima's government gave priority to integrating foreigners who had already settled in Austria rather than to further immigration.

Another demand of the FPÖ had already been taken over en bloc by the government in its coalition pact. This involved the liquidation of the social agreement with the successor states of former Yugoslavia and with Turkey and Tunisia. This meant that family allowances for children of foreign workers in their native countries were abolished. This was a long demand of the FPÖ and as such had been attacked as xenophobic and inhumane. Now this was one of the cut-back measures of the new coalition. Vienna's SPÖ mayor Häupl in an interview with the magazine *Profil*, 17 June 1996, remarked that payments had gone to children who did not even exist and "for something which I cannot control in Turkey, I don't pay up." The dilemma for the coalition was that although it adopted piecemeal many proposals of Haider, it could never satisfy his followers and could only succeed in alienating liberals in its own ranks.

TABLE 5
POPULAR INITIATIVES IN AUSTRIA, 1945-1997

Year	Topic	Signatures	% of those eligible
1964	ORF Broadcasting Reform	832,353	17.27
1969	For a 40 hour week	889,659	17.74
1969	Abolition of 13th School year	339,407	6.77
1975	Protection of human life	895,665	17.93
1980	Pro Zwentendorf	421,282	8.04
	Anti Zwentendorf	147,016	2.80
1982	Opposed to a conference centre	1,361,562	25.74
1985	Konrad Lorenz Initiative for the Environment	353,906	6.55
1985	For an extension of non-military service	196,376	3.63
1985	For a referendum on fighter jets	121,182	2.23
1986	Anti fighter jets (in Styria only)	244,254	4.50 (28.6 in Styria)
1987	Anti Political Privileges (FPÖ)	250,697	4.57
1989	For a reduction in class sizes	219,127	3.93
1989	For Broadcasting Freedom in Austria (FPÖ)	109,197	1.95
1991	For a referendum on the European Economic Area	126,834	2.25
1993	Austria First (FPÖ)	416,531	7.35
1995	Pro Motor Bike Initiative	75,525	1.31
1996	Animal Protection Initiative	459,096	7.96
1996	For Austrian Neutrality	358,156	6.21
1997	Anti genetically-manipulated organisms	1,225,790	21.23
1997	For Women's Rights	644,665	11.17

Source: Ministry of Interior, Vienna.

AUSTRIA FIRST
12 Points of the Popular Initiative 1993

1. *A constitutional provision: "Austria is no country of immigration."*
   On account of its size and density of population, Austria is no country of immigration. Whereas on average in Europe there are 100 inhabitants per square kilometre of settled land, this amounts to 230 inhabitants in Austria.

2. *An end to immigration until a satisfactory solution to the problem of illegal foreigners has been found, and until the accommodation shortage has been resolved and unemployment is down to five per cent.*
   In Vienna about 100,000 foreigners live illegally. This puts extra pressure on the labour market and accommodation. Only by an end to immigration can further social conflicts between the indigenous population and foreigners be prevented.

3. *An ID requirement for foreign employees at the work place which should be presented for the work permit and for registration for health insurance.*
   Only controls can put a stop to the illegal hiring of foreigners, which has meant not only tax evasion and the by-passing of compulsory social insurance contributions, but has also led, through cheap labour, to a decline in wage levels. The need for an appropriate regulation was acknowledged by the government in its 1990 programme but it now rejects its implementation.

4. *An expansion of the police force (aliens and criminal branches) as well as better pay and resources to trace illegal foreigners and to effectively combat crime, especially organised crime.*
   To be effective it is necessary to increase manpower. This can only be achieved through making the profession more attractive. In the first instance this includes an increase in pay and adequate provision of a modern infrastructure.

5. *Immediate creation of permanent border controls (customs police) in place of the army.*
   The auxiliary employment of the army on Austria's borders has become a long term feature. The creation of a separate border patrol from the police and customs officials is absolutely vital.

6. *A reduction of tension in schools by limiting the percentage of pupils with a foreign mother tongue in elementary and vocational schools to a maximum of 30 per cent; in case of more than 30 per cent of foreign speaking children, special classes for foreigners should be set up.*
The preservation of our cultural identity, the achievement of educational goals and the need for integration all make a limitation on the percentage of foreign speaking children in classes indispensable.

7. *Reduction of tension in schools through participation in regular education by those with an adequate knowledge of German.*
In preparatory classes children of school age with a foreign mother tongue should be taught German in order to enable them to take part in education in regular school classes.

8. *No right to vote for foreigners in general elections.*
The opposite demand of the coalition government and the Greens is primarily aimed at new votes which seek to compensate for their recent losses.

9. *No premature granting of Austrian citizenship.*
We demand that the ten year period laid down in the law should be adhered to and exceptions should be kept to an absolute minimum.

10. *Rigorous measures against illegal business activities of foreigners and the abuse of social benefits.*
Many associations of foreigners run restaurants and clubs which do not meet commercial, health or legal requirements. Some serve as centres for the black market.

11. *Immediate deportation and residence ban for foreign offenders of the law.*
The crime rate among foreigners, especially in Vienna, has soared making it necessary to provide extra detention cells. In practice deportees cannot be detained because of the acute lack of cells.

12. *The establishment of an Eastern Europe Foundation to prevent migration.*
The lasting improvement of conditions of life in Eastern European countries should be provided by specially targeted economic help to prevent emigration for economic reasons.

NOTES

1. "Popular Initiative"
To start a popular initiative support is required from either 10,000 registered voters or eight members of the lower house of parliament (*Nationalrat*), or four deputies in the provincial assemblies from three provinces. All those normally entitled to vote can sign an initiative during the period of one week by going to official booths.
If at least 100,000 sign then the lower house of parliament must deal with the issue. A committee must start deliberations within a month and has another 6 months to report to parliament. Apart from this there is no other obligation for parliament to respond to an initiative.
2. Schmidt had for some time been out of tune with the FPÖ. She had opposed the alleged rabid "blood and soil" ideology by some in the Carinthian party and was unhappy with Haider's reference to the Nazis' ordered employment policies. In 1992 she was worried by the growing influence of Andreas Mölzer, chief ideological advisor to Haider. Mölzer, a pan-German nationalist, had warned against an impending exchange of the population in Austria resulting from immigration. Haider stuck by Mölzer who was in charge of the research and educational work, on the principle of "whoever is against Mölzer is against me."
In 1992 Schmidt was the party's candidate for the presidential election but suffered a humiliating temporary withdrawal of support from Haider during the campaign after a difference of opinion. Schmidt was critical of what she perceived to be Haider's "zig zag" course on Europe and became dismayed by his increasing "autocratic" and "demagogic" leadership style. She became increasingly isolated in the FPÖ after other liberal colleagues withdrew from political activity in the party.
3. See *Neue Freie Zeitung*, 24 May 1995.
4. Interview in *Profil*, June 1995.
5. Cited in *Die Presse*, 10 June 1996.
6. Voluntary Resettlement: Speech (abridged) by Bernie Grant (Labour), House of Commons, London, 19.12.95

Madam Speaker, in principle there is nothing new about the British Government providing financial assistance to people, including British citizens, who wish to resettle abroad.

For many years there have existed various schemes. Until 1988, for example, the old supplementary benefits system provided the full costs of fares for applicants and their dependents, who wished to resettle abroad, to do so.

Since 1971 however there has also been very limited provision available under the Immigration Act of that year. I refer to the scheme run on behalf of the government by international social services and located in Brixton. This scheme provides for those who are not British citizens to obtain the cost of fares for themselves and their families plus a small amount for the transportation of their effects.

In addition to these schemes there are others, at the European level, run by the International Organisation for Migration (IOM) which facilitate return to the country of origin. The IOM's excellent programme for the return and reintegration of qualified African nationals began in 1983. The project is now paid for by the European Union.

Interestingly the IOM also ran a programme last year to assist 400 professionals to return to Jamaica, and it is hoped to expand this scheme to the whole of the Caribbean in due course. I have been told by the co-ordinator, that "this programme has aroused a great deal of interest and from our conversations with applicants we know there are many who wish to return to Jamaica."

Madam Speaker, it is now time for there to be a thorough review of existing provision in this area and for a more realistic scheme to facilitate resettlement. It is my considered view that this would be in the interests of Britain, in the interests of the Caribbean and most certainly in the interests of those who wish to resettle.

There is no doubt that there is increased interest in resettlement. Those who have already settled, include both those originally born in the region as well as those born in Britain. However they are the more prosperous, retired or professionals with marketable skills, or entrepreneurs, who can afford the travel and resettlement costs involved. There are many others who would dearly wish to resettle but who do not have the means to do so. They include the elderly, many of whom came here in the 1950s and 60s, at the specific request of the British government and never intended to remain here for long. There are others who are chronically ill and never likely to work again in the UK and who would prefer to return home. Like the retired, they are frequently trapped by a benefit system which would cut them off from benefits which they have earned if they were to go back. Given the high proportion of unemployed black workers who are long term unemployed, it makes little sense for the social security system to close off the option of return.

The advantages to Britain of such a scheme would be numerous. First of all, it is not generally in the interests of any country to have a substantial number of people in its midst who honestly do not want to be in that country. Secondly even with the most generous scheme there would be substantial financial advantages. Those who approach me are all too often those who are most dependent on social benefits, health and welfare services and indeed are most likely to be subsidised in various ways for their housing costs.

Finally an enhanced resettlement programme would be in the interests of Britain, not least because it would comprise a recognition of the huge damage which it has done to the Caribbean, both now and in the past.

Some have said that even to mention this matter, is to cause damage to race relations in this country, and to argue for a resettlement scheme is to give in to racism. I have even been accused of adopting the agenda of fascists and racists. Even the Secretary of State for Home Affairs told me recently that he feared that an enhanced resettlement scheme would make black people feel unwelcome here.

I am not convinced by these arguments. This is about creating positive choices for black people, setting our own agenda for once, and about remembering where we came from. Some black people will want to stay in Britain and the fight for racial equality will continue. I myself will continue to play a part in that struggle. It is true that some want to leave because they are sick and tired of the racism they face in this country. There are others however who quite honestly do not feel at home here and others who despair of the predominant values and culture of this society. Then there are others who quite simply wish to return to live with their families.

This House has no right to deny those people a choice about their future.

Relaxing
(Photo: Holzner)

**1974 Press conference, as leader of Free Youth of Austria
(Photo: Fritz Kern)**

**Election as leader of FPÖ, Innsbruck 1986
(Photo: Holzner)**

With Franz Vranitzky
(Photo: Holzner)

Voting with his wife
(Photo: Holzner)

With an estranged Heide Schmidt
(Photo: Holzner)

With President Kurt Waldheim
(Photo: Holzner)

Top: With Niki Lauda, racing driver and airline owner
Bottom: The FPÖ moves towards the Church (Photos: Fritz)

**Top: Thousands demonstrate in Klagenfurt in protest against Haider's dismissal as provincial governor in 1991 (Photo: Fritz)
Bottom: Addressing the Ulrichsberg Festival (Photo: Eggenberger)**

Top: Rock climbing (Photo: Fritz)
Bottom: Victorious at sea (Photo: Eggenberger)

Ewald Stadler, Haider's number 2 in parliament; nicknamed "the Doberman" (Photo: Holzner)

Looking into the future (Photo: Fritz)

8

EUROSCEPTICISM

After the result of the petition drive on immigration Haider sharpened his critical position on Europe. He looked up the party programme of 1985 and concluded it was incompatible with the Maastricht project which envisaged a centralised Europe. In an information booklet on Europe produced by the party in 1993 a clear preference was discernible for a "confederation [against a United States of Europe]."[1] Less than a year before in the Vienna Declaration Haider had described his position thus: "Whoever goes with me stands for an FPÖ which supports a United States of Europe as the lasting guarantee for peace, and whose basis will be the right to self-determination of peoples as well as the human right to a homeland."[2] Nuances such as these led some liberals in the FPÖ to break with the party. They were dismayed by its position on immigration and Europe and founded a Liberal Forum. Once regarded as the party for European integration, the FPÖ under Haider became increasingly sceptical of the whole project in the way it was unfolding. Whilst he could go along with a confederation of sovereign nations, the FPÖ leader rejected the idea of a single centralised European state.

When it came in June 1994 to taking a stance in the referendum on joining the EU, the party opted for a "no" vote. The Austrian people thought differently and returned an overwhelming majority in favour of the coalition government's negotiation package and for Europe (see Table 6). Like the petition drive on immigration, the referendum result was regarded as a massive setback for the Freedom Party.

Haider outlined his position on Europe in his book, *The Freedom I Mean*. It closely followed the general Eurosceptic view as expressed by the Conservative Right in Britain and by people like Manfred Brunner in Germany. This was fiercely anti-Maastricht but not necessarily anti-Europe. They believed that Europe could only be saved if built in a different way from that envisaged under Maastricht. They considered that the Euro-elites had drifted apart from their peoples who felt increasingly alienated from anonymous institutions. The message of referenda on Europe such as those held in Denmark and France indicated a strong undercurrent against the push for unity at all costs. In Britain people like Margaret Thatcher cautioned against building a Europe without the people and energetically called for something to be done to address the "democratic deficit." Haider took up this theme in Austria:

> Solving the lack of democratic procedures remains on the long agenda of overdue reforms. But the Euro-fanatics continue to accelerate the pace of integration regardless of the cost to democracy or the people. [Maastricht was] one treaty too many and one step too far.... Anyone who has a vision of freedom for Europe should know that the new democracies of Eastern Europe have a profound distaste for the spectre of centralism. It was not for nothing that they threw off the stifling mantle of a jaded bureaucratic superstructure. Maastricht is the wrong signal for today's Europe. A Europe which is divorced from its citizens is unthinkable and will go the way of despotic empires before it.[3]

Haider railed against the "Maastricht superstate" with its "faceless bureaucrats" and stoutly defended the interests of the nation states against "meddling and interference" from Brussels. For him, like Margaret Thatcher, it was clear that "seventy years of the Soviet Union show the futility of trying to steamroller freedoms and national diversity. Political imperialism and colonialism cannot stamp out cultural identities of peoples.... The nation state is not finished. To try to abolish it is the quickest path to nationalism." Haider concluded that, "All previous efforts to establish a super state ended in the garbage bins of history. By clamping down on natural sentiments and feelings, chaos is provoked, because eventually the desires for freedom will explode." The alternative was to be a Europe "from the Atlantic to the Urals" resembling the vision of the former French president General De Gaulle.

Haider shared the modern Eurosceptic view that harmonisation in the EU was insane and extreme. "With one touch of a computer key in Brussels local identities can be extinguished, offending dialects erased and vegetables harmonised." He supported the view of the Conservative Michael Spicer, a leading British Eurosceptic, who argued that, "the future of Europe lies not in the formation of a new single state but in the close association of old nations co-operating with each other in a wide variety of ways." This was not the mainstream thought in Austria and many in the ÖVP looked forward to a fully-fledged federal Europe with common monetary, foreign and defence policies.

Haider detected similarities between problems in Austria and in Europe as a whole. For him there was evidence of a general malaise and a feeling of unrest amongst the peoples of Europe with their leaders. This frustration was periodically registered in elections by a large protest vote or high abstentions. It was symbolic of a gulf between the governed and those in government. In the EU this disenchantment was magnified: "Throughout Europe there is a general loss of confidence in the political establishment. In the EU bureaucracy as in Austria, the ruling parties share out jobs, pensions and pay offs neatly between them." Haider concluded with Michael Portillo, darling of the Eurosceptics in Britain, that it "was time to stop the rot in Europe." He argued for a "contract for Europe," for more democracy, for the sovereignty of the member states and for more freedom. Haider saw the EU plagued by the same problems which he disliked in Austria – bureaucratic red tape, overcentralisation and over-regulation, corruption and patronage.

Haider resolutely opposed the drive to establish a federal Europe and argued instead for a looser intergovernmental cooperation.

> This is a model which will bring more rights of self-determination and freedom to the peoples of Europe, than some arbitrary centralised supra-national body. Only the nation state can protect the heritage and culture of a people, a treasure which should be kept.... Maastricht is dead. Not just because of British and Danish scepticism and various opt-out clauses. Not just because the currency union is unattainable. Not just because the German constitutional court formulated substantial legal reservations on October 12 1993. Maastricht is dead because it was never viable from the start but was stillborn.

Haider believed his reservations on Europe emanated from a love of Austria. He maintained that because of its position on immigration and Europe, the FPÖ could be considered the only real "pro-Austria" party. This sentiment had been expressed in a resolution at a special party conference 8 May 1993, published in the party's paper, *Neue Freie Zeitung*, 12 May 1993, under the title "Austria first – our way to Europe. For a Europe of the citizens and peoples. For freedom, peace and prosperity." This stressed the importance of decentralisation and a confederation of national states and regions working on the principle of subsidiarity.

The party saw radical constitutional reform as a precondition for entry to the European Union. Part of this was to include an upgrade for the second chamber in parliament, the federal council, which had little clout in the decision-making process (see Appendix 1, Figure A1). Many, especially in the ÖVP, appreciated the need for a thorough reform of the weak federalist model as applied in Austria. It had been hoped that a federal reform bill would accompany entry to Europe. This aim was postponed after the 1994 election when the two coalition parties lost the two thirds majority necessary to pass important constitutional legislation.

In a publication issued by the FPÖ research department in 1993, *Der Weg nach Europa*, the case was made for constitutional reform. This included a call for the direct election of town mayors and provincial governors, and a greater role for the instrument of direct democracy to check bureaucracy and maintain contact with the people. Before the referendum on the EU, the FPÖ concluded that the government had not done the necessary preparatory homework to justify its support for membership. A resolution of the party in April 1994 reiterated its adherence to the process of European integration and an all-European confederation on the basis of democracy, federalism, the rule of law and full preservation of the rights of self-determination. Further the party called for tax reforms, generous compensation for farmers, guarantees on jobs and for clarification of the country's position on security issues. It concluded that negligence on the part of the government to sufficiently safeguard Austrian interests forced the party to go for a "no" vote on EU membership. Not all in the party were happy with this decision. The business wing feared the consequences for Austrian industry if the country stayed out and many others had long advocated joining Europe. The EU tactic was a high risk gamble which tried to square a pro-European circle with scepticism. At a special congress in Villach on the EU in April 1994 Haider after listening to arguments for and against concluded he was for a future in Europe but for one in which the citizens had the say and not the "Apparatschiks in Brussels."

Despite claims by the FPÖ to represent the interests of the Austrian people, the campaign for the EU referendum did not go according to plan (see Tables 6 and 7). The result was a disappointment for the FPÖ which tried to shrug off the result as the product of one-sided government propaganda. Certainly the party once again faced all the big guns of the Second Republic. From the federal president down, including the coalition parties, the social partners, and most of the media the mood was passionately for a "yes" vote on Europe. The FPÖ stood out against this barrage, not completely isolated but in the strange company of the Greens and some on the outer left. As in Britain an unholy alliance emerged between the far right and left in joint opposition to the EU Europe.

A weakness of the "no" campaign was that it was uncoordinated and prone to hyperbole. Scare stories of what would happen to Austria in the European club were wildly exaggerated and damaged the credibility of the "no" camp. Haider contributed in his own unique way by presenting a tub of Spanish yoghurt on television, claiming it to be infested with insects.[4] This was easily refuted by Haider's opponents who took some delight in ridiculing his knowledge on EU food standards and the labelling of products.

Their glee was to be short-lived and revenge for Haider sweet. A few months later Haider's party reversed its fortunes in the parliamentary election. In the long term opposition to Europe could be expected to pay dividends. The government had been forced to make generous promises before the referendum to ensure a positive vote. Chancellor Vranitzky had made a fatal pledge not to raise taxes because of the EU, a position which became increasingly untenable with the need to balance the budget deficit. During the EU campaign reassurances had been made by the government to farmers on prices and to the population in western Austria on transit traffic, in an effort to secure their support. Expectations of Europe were high: jobs would appear and goods would be cheaper. It was inevitable that this euphoria would subside and just six months after entry to the EU polls showed increasing scepticism on Europe. Haider could feel his position vindicated and looked forward to a new source of protest votes from disaffected Europeanists.

One year after the referendum Haider launched an attack on the coalition's European policy. He attacked the enterprise as an unacceptable burden to the Austrian taxpayer who had to fork out 100 million Schillings a day for the privilege of membership. Six months later, Haider realised even he had made a mistake and that the daily cost was in reality 150 million. Haider wanted to know where the money went and suspected much got lost in the dubious channels of the EU notorious for fraud and lack of controls. Much like Margaret Thatcher before him, Haider demanded a rebate from

Brussels. Austria was a hefty net contributor to EU funds and now Haider was looking for ways to "get this money back." He also appeared as the protector of many farmers who faced financial difficulties because of membership. Many were entitled to compensation but procedures were long and bureaucratic and benefits were limited in duration. As far as the "yoghurt fiasco" was concerned, Haider felt his position justified and quoted from a booklet produced by the Chamber of Labour on labelling of foodstuffs. According to this some products in the EU contained additives and flavouring which were damaging to health. These were marked with obscure codes not immediately obvious to shoppers in the supermarket without the right reference book at hand. Haider made the most of the fact that prices did not come down as expected and that transit traffic in the Alpine regions became more of a problem and not less. It all seemed to justify his argument that the government had carried through the negotiations with too much haste and had not done the sufficient "homework" before entry.

The FPÖ had urged structural reforms in the Austrian economy before entry to avoid a sell off of Austrian economic interests and to prevent a budgetary catastrophe. These remedial measures included a strengthening of capital resources and a reduction in secondary wage costs. A fundamental overhaul in the tax system and structural reform in the banking and insurance sector were also needed. Haider and the Freedom Party further stressed the need to de-politicise the National Bank (the Federal Reserve Bank), dominated by the Reds and the Blacks.

The government called for patience from the population arguing that benefits of EU membership would take time to filter down. But time was not on the side of the coalition and people were fast running out of patience. By the end of 1994 the year of the referendum it was plain that the government's policy on Europe was drifting. It was unable to formulate a coherent, identifiable position on Austria's role in Europe or the future of neutrality in a new defence and security system. The two parties were often at loggerheads and could be seen squabbling for plum posts in Brussels. Haider made the most of this confusion and loss of momentum. One area he asserted, where the government had not been lax was in allocating good jobs in Brussels for their own people, who were not always the best qualified candidates but simply loyal party has-beens looking for a soft option. When it came to protecting their own privileges and interests, he added, the old parties were European champions. At the same time people were having to accept cut-backs and many less well-off families were confronted with a reduction in family allowances as well as increases in gas and electricity charges as a result of a so-called "ecological tax."

In addition the government had not bargained for a weaker Lira in Italy. This, combined with unequal tax rates, led to a regular exodus of shoppers to neighbouring Italy. Bargain hunters using the freedom of the internal market could enjoy lavish shopping sprees there which hit industry particularly in provinces such as Carinthia on the border. According to Haider this all served to undermine the Austrian economy and jeopardise jobs. On the issue of the EU as on immigration, Haider promoted his image as the protector of Austrian interests, culture and traditions.

Austria joined the European Union on January 1 1995 but there was no sign of the necessary direct elections for the European Parliament in Strasbourg. The government consistently dodged the question weakly replying that procedures were long and complicated. By the end of 1995 the Swedes had managed to hold their Euro-election which confirmed that widespread disenchantment with the EU was not confined to Austria alone. The Austrian coalition could not have been encouraged by this result. It feared a large protest vote would benefit Haider and stalled. The coalition staggered on from one crisis to another and finally ground to a halt at the end of 1995 paralysed by internal wrangling over the budget.

In the long term Europe was a vote-getter for Haider and the FPÖ. His position was in tune with the popular mood and growing Euroscepticism elsewhere. This was an example of Haider's political astuteness and talent for judging the right time for policy shifts and adjustments which in the short term could not be expected to yield much political capital. This was a quality which his former colleagues however did not rate as one of Haider's good points. They cited it as crude opportunism which looked at politics as a way of achieving maximum gain.[5] In this exercise principle and morals were sacrificed to the sole end of the electoral well-being of the party.

Europe illustrated the dilemma in analysing the Haider phenomenon. Was he a "normal" smart politician, a smooth operator with a feel for the game or an unprincipled demagogue out to get as much for himself at whatever cost? Did he have a genuine desire to renew Austria and shake it out of its stone-age lethargy giving people more say over their own future free from the Reds and the Blacks? Or was he interested in building up his own power cartel where he could settle old accounts and establish personal rule? Many politicians, if at all successful, switched tactics and policy to respond to the public. The SPÖ had also shifted ground on Europe and neutrality as well as domestic issues such as immigration. Very often thoughts of the next election prompted an updated version of party policy. The whole exercise could be justified as democratic with the comment that the party was listening to the people "out there." Many politicians pushed through internally in the party their own proposals and rode roughshod over opponents. The poor image of politicians in the modern world has conditioned many voters to expect this of their representatives. For the cynic Haider was a clever politician, whereas for the purist he was unscrupulous and potentially dangerous.

SECURITY

On the question of Europe Haider had been accused of crass opportunism. He had, it was said, transformed the FPÖ, one of the early converts to Europe into a bunch of rabid sceptics purely for the maximisation of votes. This approach to politics seemed to fit the brash image of a wily politician determined to get to the top of the "greasy pole" yet it belied a complexity which was apparent in other areas of policy. On some issues, Haider went against the current and defied the tide of popular opinion. One of these was neutrality and another the question of NATO membership.

Austrian neutrality was, like social partnership, a holy cow of the Second Republic. People were fond of it and believed it offered guarantees against all possible evils from communism to nuclear fallout. The "island of the blessed" felt secure with the girth of neutrality around it, a shield however which looked ever more vulnerable in the post cold war world. Neutrality had been adopted in 1955 by the Austrian parliament and was expressed in a constitutional law of 26 October of that year which stipulated that:

> 1. For the purpose of the permanent maintenance of her external independence and for the purpose of the inviolability of her territory, Austria of her own free will declares herewith her permanent neutrality which she is resolved to maintain and defend with all the means at her disposal.

2. In order to secure these purposes Austria will never in the future accede to any military alliances nor permit the establishment of military bases of foreign states on her territory.

Austrian politicians had generally dodged the issue when it came to an open and frank discussion of neutrality. They preferred to hedge their bets speaking vaguely of a new security architecture which would emerge sometime in the future until which time neutrality should remain. This was the stock response of Franz Vranitzky as chancellor and many social democrats who feared any other course would lead to militarism. The ÖVP acknowledged the changing realities in the new world and had less of a complex about joining the WEU or even a revamped NATO. Along with President Klestil the ÖVP people liked to substitute the word solidarity for neutrality. Its leaders also considered there could still be a residual role for neutrality in areas other than Europe, for example in Africa or South America.

The coalition pact of March 1996 between the SPÖ and ÖVP referred to the need to examine full membership of the WEU in the light of developments at the EU inter-governmental conference. A report was to go before the Austrian parliament by the end of March 1998 at the latest. On the basis of this, the government would put proposals necessary before parliament for debate. NATO was not explicitly mentioned and neutrality omitted. The coalition pact instead referred to the need to dynamically develop Austria's relationship with other security organisations operating within the framework of the EU member countries.

The debate on neutrality straggled on with the SPÖ bringing up the rear but steadily modifying its position to accept that neutrality was not a God-given reference point.[6] Gradually even members of the SPÖ realised security had a price tag and began to reformulate their views on neutrality as well as NATO membership in various press interviews. The result was a string of ambiguous statements which in the end created the impression of division within the SPÖ as well as the coalition. During the debate on the EU ratification in parliament in 1994, the SPÖ talked of neutrality as an integral part of government policy whilst the ÖVP thought it had been useful but belonged in a museum. The continual fudge on security, neutrality and NATO membership confused many and a coherent government position remained elusive.

The FPÖ had less problems with the entire security issue. As early as September 1990, Haider had suggested that the question of neutrality should be opened up for discussion in the light of changes in Eastern Europe. The media and the coalition politicians immediately jumped on Haider as if he had committed high treason. This time he could not be accused of populism since polls indicated an overwhelming support for neutrality. If anything this charge could be levelled at the Socialists with their peculiar nostalgia for the lost age of neutrality. As so often when Haider made a thoughtful comment it came with a sour side. His contribution to a sensitive debate was made not in Austria but in Munich in neighbouring Germany, the haunt as his adversaries were only too keen to point out of Hitler and his henchmen. This was an unfortunate place for such a policy statement but somehow typical for Haider: it took him one step forward but two steps back and removed the discussion on neutrality from the bounds of objectivity.

The FPÖ unequivocally argued for NATO membership and more money for defence to build up a professional army to cope with future commitments. Austria only spent 0.9 percent of the GDP on defence and its armed forces were inadequately equipped to deal with extra obligations. Gradually Austria was drifting to closer if imprecise co-operation with its partners. It had observer status in the WEU and had joined the NATO partnership for peace. A quantum leap to WEU and/or NATO membership had consequences for Austria's status as a neutral country and to abandon neutrality technically required a two-thirds majority in parliament. Experts disagreed on whether neutrality as defined in the federal constitutional law of 1955 was compatible with Austria's new role in Europe.[7] Arguably Austria would not be joining a "military alliance" but a defence community and would not be obliged to allow "foreign" military bases on her soil. The matter was further complicated by the ongoing discussion in Europe on the changing role of NATO and new approaches to security. The backdrop to the discussion was constantly moving which did not help people to focus on the problem. On this question at least, the FPÖ was prepared to lead from the front regardless of electoral implications.

ELECTION '94

The electoral prospects for the FPÖ looked distinctly favourable and were confirmed by substantial gains in the 1994 national council election (see Table 1 and Figure 5). The SPÖ lost 15 seats, and the ÖVP lost 8 giving it only 10 more seats in the chamber than the FPÖ.

These results came as a shock to the governing parties who had scored such a big success with the EU referendum only a few months earlier. The good will and co-operation between ÖVP and SPÖ which had been apparent during the referendum campaign disappeared. The Haider factor once again could operate to the full. Chancellor Vranitzky looked vulnerable on television particularly against professional "power women" politicians from the opposition. The Greens and the Liberals exposed what they saw to be weaknesses in the SPÖ's social programme. Both attacked the SPÖ which had been the "chancellor party" for nearly a quarter of a century but had, so they argued, neglected the less well off and the position of women in society.

Haider ran a typically effusive campaign with simple slogans such as "Simply honest, simply Jörg." He attacked with devastating effect the "sleaze" factor in a system riddled with party politics and patronage. A prime target was the socialist dominated Chamber of Labour whose leading members enjoyed exorbitant salaries and generous pensions beyond the wildest dreams of the workers they were supposed to represent. Elections were pending too in the Chamber of Labour and here the FPÖ list increased its share of the vote from 7.7 percent to 14.4 percent. A further reward for Haider came with the subsequent national council election which brought gains amongst the blue-collar workers (see also Tables 8 and 9) and the support of over a million voters.[8] This election was a breakthrough for the FPÖ and confirmed it as one of the most successful right of centre parties in Europe.

Once again the SPÖ went into a coalition with the ÖVP but the new government lacked confidence and was burdened with the job of introducing unpopular cut backs in social welfare. Socialists tried to sell this policy as "correctives" to the welfare state but those affected feared its demolition. In this climate Robin Hood-Haider could only flourish and wait for further recruits from the socialist camp. The 1994 election was one of the great highlights for the FPÖ and for

Haider personally who had dominated the campaign so brilliantly on television. He was believed by his adulating followers and bitter opponents alike to be unstoppable. His self-professed goal – to move into the chancellory at the famous Vienna Ballhausplatz – had moved substantially closer. No other politician provoked such strong emotions whether positive or negative from Austrians as Jörg Haider. The choice was simple – one could be for or against the Freedom Party leader but not neutral. Inevitably this sensitivity made an "objective" not to mention "fair" assessment of Haider complicated and open to misinterpretation.

TABLE 6
RESULTS OF THE EU REFERENDUM, 1994

	Valid Votes			
Electoral Region	Yes	%	No	%
Austria	3,145,981	66.58	1,578,850	33.42
Burgenland	148,041	74.66	50,238	25.34
Carinthia	232,457	68.20	108,410	31.80
Lower Austria	678,988	67.93	320,483	32.07
Salzburg	184,948	65.06	99,335	34.94
Styria	501,481	68.88	226,556	31.12
Tyrol	198,990	56.66	152,211	43.34
Upper Austria	539,965	65.49	284,547	34.51
Vienna	542,905	66.15	277,770	33.85
Vorarlberg	118,206	66.59	59,300	33.41

Source: *Wiener Zeitung*, 24.6.1994.

TABLE 7
EU REFERENDUM VOTING BEHAVIOUR (%)

		Yes	No	Difference
Men		70	30	+40
Women		62	38	+24
Aged	under 30	55	45	+10
	between 30 and 44	64	35	+29
	between 45 and 59	68	32	+36
	over 60	70	30	+40
Self-employed		63	37	+26
Farmers		30	70	-40
Salaried employees/civil servants		67	33	+34
Blue-collar workers		64	36	+28
Male pensioners		68	32	+36
Female pensioners		63	37	+26
Housewives		61	39	+22
In education / not employed		61	39	+22

Source: A. Pelinka (ed), *EU Referendum*, Signum, Vienna, 1994, p. 97.

NOTES

1. *Der Weg nach Europa: die Hausaufgaben der österreichischen Bundesregierung aus freiheitlicher Sicht* (Vienna: FPÖ, Bildungswerk, 1993), p.8.
2. *Wiener Erklärung*, speech by Jörg Haider, 7 April 1992, published by the FPÖ, "Wer mit mir geht, steht für eine FPÖ, die für die Vereinigten Staaten von Europa als dauerhafte Friedensgarantie eintritt, deren Grundlage das Selbstbestimmungsrecht der Völker sowie das Menschenrecht auf Heimat ist," p. 63.
3. Quotations from J. Haider, *The Freedom I Mean* (New York: Swan, 1995), pp. 81-83.
4. See M. Sully, "The Austrian Referendum 1994," *Electoral Studies*, vol. 14, no. 1, March 1995, pp. 67-69.
5. Interview with Helmut Peter, former FPÖ member, and now member of parliament for the Liberals, July 1996. He wrote the following open letter (abridged) to Haider when he broke with the FPÖ:

From Helmut Peter to Jörg Haider, 4 August 1993

The Freedom Party has, under your leadership, been given an undreamt of boost. Your energy and political skill to open up issues have noticeably changed the republic of Austria. The dual division of the country into a red and a black sphere of influence still exists as before but the rulers have been forced to become more prudent since you appeared as a controlling force. Seen in this way your work in opposition is successful. Your personality has become a distinctive fixed point of the political scenery.

On your political path you have fundamentally changed the FPÖ and moved it from a middle-class position as a national-liberal party to the right, conservative edge of the party spectrum. There you successfully collect votes of people whose emotions you inflame and intensify, without any consideration for the thoughts you evoke.

You are right: votes in a democracy are not weighed but counted. The respect for every opinion of a voter is the precondition for parliamentarism. You forget in this all too easily, that the representatives of the people are there not only to strengthen public opinion or groups in the population, but have instructional, informative and even educational tasks vis-à-vis the voters. Not the short term wish of voters but the long term good of the community must stay in the foreground.

To grasp the worries of people, to go to them, to listen to them and to try to understand them is one side of the coin. To incite feelings of distress, to reinforce and stir up potential fears, is the other side. I consider that irresponsible. It makes no contribution to the necessary solution of problems but even makes this impossible.

To put this weak government constantly under oppositional pressure and to conduct tough discussion is the right way. Abuse and personal insults of the representatives and decision makers of our country is bad form which allows political rivalry to degenerate into open enmity.

The behaviour of political rivals cannot justify one's own verbal radicalism.

The integration of Austria in a common Europe, in the form of joining the EC was a programmatic foundation of the FPÖ: You have been very successful here in converting a party of European supporters into one of critics and opponents of Europe. This seems to you useful in terms of electoral tactics and probably you are right.

Agreed the government is here also weak and hesitant and here there are many good oppositional points for attack and room for hard criticism.... I still believe today that you personally consider EC integration to be meaningful. But the chance to get hold of votes from the reservoir of the EC opponents, even with the risk that the EC vote will have a negative outcome, is more important for you. That may well be party politically opportune but has nothing to do with a responsible position.

It's very easy to come out against the structural, organisational and democratic political shortcomings of the EC and to polemicise with verve. It's much more difficult however to answer the question how we can solve the problems of our country in the present and future without EC membership.

You are not prepared to carry out this work of convincing people but prefer through agitation to make people feel insecure and in this way endanger the success of membership.

You have become one of the best known and most powerful people in Austria but at the same time perhaps one of the most lonely. Your energetic but unfortunately immoderate politics has made you the lonesome supreme ruler of the FPÖ, which has lost internal, critical potential.

It is one of your greatest personal deficits that you only understand how to shine as solitary fighter but you are incapable of team work. You have lost personalities through this who were always prepared to compromise and be loyal and who long stood by you. The authoritarian leadership principle is long since obsolete in politics.

So you have become the single thinker of your party. You have collected many personalities around you who are used to bless and implement afterwards what you lay down. Naturally it is discussed openly and everything is democratically decided but a process of opinion formation doesn't take place any more.

Your brilliant but often unrestrained and hurtful rhetoric has cost you the esteem of your political enemies. The controversy the "FPÖ versus the rest of the world" cannot be won and leads despite all election victories into political isolation.

The international reputation of your FPÖ, and it is now only your party, is destroyed.

The mere rhetorical disassociation from right extremism and its terrible consequences in history are too little, it must be felt deeply and be credible.

In all hard political debate, I try to find common ground rather than differences.

Political opponents are not the same as enemies.

This may sound old-fashioned but it is simply my way.

I am therefore with effect from today resigning from the Freedom Party.

6. See interview with Josef Cap, SPÖ member of parliament and former ardent opponent of NATO in *Profil*, "Kurzfristig in die Nato," 15 July 1996. Cap argued for an open discussion on NATO membership and an end to taboos on a professional army and the end of neutrality. The interview triggered an extensive debate on NATO with different SPÖ politicians making different individual statements.

7. See article in *Die Presse* by W. Pahr (former Foreign Minister) and F. Cede (member of the Austrian Foreign Ministry), "De-facto-Ende für Österreichs Neutralität," 27 June 1996. See also J. Haider, *Friede durch Sicherheit* (Vienna: Freiheitliche Akademie, 1996), pp. 142ff.

8. See M. Sully, "Austria's New Government – Fresh Start or Swan Song?" *The World Today*, February 1995, pp. 24-26.

9

ELECTION '95

Despite their losses in October 1994 (see Figure 5), the SPÖ and the ÖVP started customary negotiations on building a coalition, albeit overshadowed by the Freedom Party breakthrough. From the beginning the exercise seemed plagued by self-doubt and fear. The leader of the SPÖ and Chancellor, Franz Vranitzky, looked jaded and beaten, and the ÖVP chief Erhard Busek was living on borrowed time until the next party conference. The government that was, in the end, cobbled together was larger than its predecessor and promises of a new start and "lean administration" soon disappeared in the November fog. In short coalition synergy was missing and the will and capacity to solve urgent problems was absent. The FPÖ revelled in the coalition's lack of self-confidence and called a number of turbulent, special parliamentary sittings. From the beginning the coalition was on the defensive and looked hopelessly adrift in a political "Bermuda triangle."

In January 1995 Austria joined the European Union with little pomp or ceremony. The coalition which had marshalled massive support for Europe in a referendum in June 1994 sensed the mood of voters had become more sober. Prices had not tumbled as predicted, a solution to the problem of heavy transit traffic in the Tyrol had failed to materialise and the Promised Land was still around the next corner. An Alpine variety of Euroscepticism had spread throughout the country and the government seemed incapable of shaking off its inertia to deal with the problem. Obsessed with further losses to the opposition, direct elections to the European Parliament in Strasbourg were delayed ostensibly for organisational and technical reasons (see interview section with Otto Habsburg). Personnel appointments to Brussels were criticised for going to the favourites of the government parties rather than being based on merit. This was noted not only by the opposition in Vienna but also by irritated officials in the European Union.

This approach to politics in Austria looked almost Byzantine in the modern world. Much of the country's neo-corporatist outlook and institutions resembled relics of a bygone age. The kind of system in which party hacks were promoted in back room deals in town hall politics astonished many foreign commentators.[1] *The Economist* remarked that:

Business men, trade unions, the central bank and so on have settled things in smoke-filled rooms while, for much of the time the reds and the blacks have shared the spoils of government. Under a system known as *Proporz*, almost all senior jobs (even in schools and hospitals, let alone embassies, town halls and banks), as well as many contracts and even public housing have been doled out by formula to suit the two big parties' supporters.[2]

The Austrian success story built around this system, social partnership and neutrality had served the country well but was beginning to look like political archaeology unsuited to the challenges of a new age. Fear of change and hypersensitivity to criticism however are deeply ingrained in the Austrian political culture.

BACKGROUND TO THE '95 ELECTION

In April 1995 a boost was given to the flagging fortunes of the not-so-great coalition and several reshuffles signalled a new start. The People's Party which had been locked in a bruising leadership crisis broke the impasse at Easter with the election of 50 year-old Wolfgang Schüssel who took over from Busek as party chairman. He also became Vice-Chancellor and Foreign Minister in the government. Relations with the Socialists were patched up and for a while even became "cozy." A rejuvenated coalition team showed a new spirit to work together. The "chemistry" between the new man Schüssel and the Chancellor Vranitzky was deemed to be positive and the ÖVP leader was reckoned to be a smart negotiator and a fan of big coalition politics. A former Economics Minister, he was respected for his ability to compromise and bargain. The bonhomie however proved to be short lived. On his election as leader at the party conference, Schüssel stated his aim to be "number one" in Austria and to become chancellor. He desperately needed to reverse the catastrophic electoral performance of his party to ensure the survival of the ÖVP, and to shore up his own position as leader. The ÖVP has not provided the chancellor in a government since 1970 and this ambition of Schüssel always looked excessively optimistic and overstretched. Schüssel knew he had to deliver the goods if he was to escape the fate of many luckless leaders of the ÖVP before him.

The rank and file in the ÖVP were becoming increasingly frustrated with playing second fiddle to the Socialists in government. Many were fed up with bearing the brunt of responsibility for the mistakes of the government, whilst the Socialists often romped off with the glory when policies went right. Others felt that after 25 years of a socialist federal chancellor it was "time for a change" and stepped up the pressure to break from the restrictions of an unpopular coalition.

A window of opportunity soon came which tempted the ÖVP to go to the country in a desperate gamble to become "number one." Throughout the summer of 1995 opinion polls showed increased sympathy for Wolfgang Schüssel. With his distinctive bow tie and large glasses, he came over as a larger than life figure on television. The party looked like it was finding some peace with itself and internal feuds were not so often leaked to the press. The new team could be seen happily enjoying the summer in folksy tradition in Alpine huts on Austria's mountains. The party also set about claiming back some of the intellectual high ground which it had yielded to the Freedom Party and took a tough line on law and order and immigration.

BUDGET TROUBLES

The main issue which tore the coalition apart was the failure to agree on a common policy to tackle the alarming budget deficit. This was likely to exceed 5 percent of the GDP in that financial year and had to be brought down to 3 percent if Austria was to qualify for membership of the single European currency. Experts agreed on the need to trim the public sector and curb social service spending. Throughout the summer of 1995 the government seemed confused on the extent of the budget deficit. Figures issued by the socialist Finance Ministry did not tally with the estimates issued by the ÖVP-run Economics Ministry. The new Finance Minister, Andreas Staribacher, was accused by the ÖVP of being naive and inexperienced and for covering up the gravity of the situation. Distrust and suspicion plagued the relations between the coalition partners who started in the autumn a round of negotiations on the budget. After weeks of fruitless bargaining it was clear that the life of the coalition was in jeopardy. Finally Vice-Chancellor Schüssel left the negotiating table convinced there was little point going round in circles.

With hindsight the SPÖ doubted whether the ÖVP had ever seriously sought an agreement and accused its leaders of putting naked ambition before the interests of the country. The Socialists claimed to detect a plot to throw the country into early elections which had been stage-managed months before by power-hungry leaders of the People's Party. For his part Schüssel complained of the obduracy of the socialist trade unionists and the incompetence of the finance minister. He nevertheless had to take much of the flak for the debacle from an electorate annoyed at what it saw as an unnecessary and costly winter election.

Schüssel went into the election promising "savings, cast iron savings" rather than tax increases to solve the budget crisis. He refused to go along with what he openly admitted would be yet another "swindle budget" which neglected to introduce fundamental structural reforms. Austrians had been living beyond their means, the limits of the welfare state had been reached and the Maastricht convergence criteria loomed on the horizon. One of the most sensitive issues was pensions' reform. It was widely acknowledged that the practice of taking early retirement in Austria was open to abuse. According to the chamber of commerce, the percentage of the population between the ages of 60 and 64 in active employment is 8.8 percent. This compares with figures of 38 percent in Britain and 19 percent in Germany. Average retirement age in Austria is 57 and in some jobs, such as on the federal railways, it is even lower. At the other end of the scale Austrian students can take up to eight years for their studies and enjoy university education free of fees. The generous welfare system further provides at the state's expense two years maternity leave. These allowances were gradually crippling the economy and were in need of review. The question was where the ax would fall and who would lose out. In this discussion it was the People's Party that was cornered into looking semi "Thatcherist" and vulnerable to the charge that the poor and the weak would be left out in the cold.

THE CAMPAIGN

The tone of the campaign can be described by the German word *Angst* which goes beyond simple fear and anxiety and verges on obsessive neurosis and near hysterical phobia. *Angst* was everywhere – in the newspapers, on television and in the minds of voters conditioned to routine consensus and solid stability. Despite its troubles Austria is still one of the richest countries of the European Union and the economy is basically in good shape. *Angst* nevertheless gripped the country as if it was on the brink of doomsday. The Socialists feared losing power, Schüssel was afraid of not getting it and everyone was afraid of Jörg Haider. Even Haider, it seemed was often unsure and less confident than of old.

For the Socialists the campaign was conducted under the inspiration of the new and "bubbly" General Secretary, Brigitta Ederer. It was decided to shield the chancellor from exposure to TV debates until the last phase of the campaign to avoid mistakes made in 1994. Labour star Tony Blair was flown in to explain the evils of any quasi-Thatcherist course and to stress the importance of social justice. The SPÖ went "back to the roots" in an effort to talk to the "little man" who in 1994 had found refuge with the FPÖ. It rediscovered the underprivileged and the "have nots" in society and pledged that it would fight for traditional working-class interests. Many in the SPÖ felt that the party had become too absorbed with power and feathering the nests of its own functionaries to care any more about the average man in the street.

The SPÖ in 1995 put forward a "chancellor programme" for social justice and for more jobs and democracy in the EU. It supported a limited but humane immigration policy, opposed the introduction of university fees and promised to tackle welfare abuse and tax dodgers. The party opposed the idea of NATO membership and promised to uphold Austrian neutrality adopted in 1955. In essence the SPÖ stood for a "conservative" policy of "safety first" and "no experiments." The chancellor's television style was analysed by experts as "lead heavy," stolid and statesmanlike. It reassured voters afraid of change, it reassured pensioners afraid of losing out and it reassured Austrians afraid of *Angst*.

The ÖVP presented the electorate with a new kind of "Raab-Kamitz" course, an exhumation of the party's 1950s economic policy which most voters had either never heard of or had forgotten. Rebaptised the "Schüssel-Ditz" course after the leader and his economics guru, the ÖVP tried to convince voters that the socialist prescription of "more of the same" was not the right medicine. Cutbacks in government expenditure could not be avoided in the view of the ÖVP to keep the currency hard and Austria competitive. The policy was unfortunately summed up by the Minister Ditz in Churchillian tones as "blood, sweat, toil and tears." This coincided with unrest and strikes in France which worried voters that such policies would provoke rioting in Austria. Both Schüssel and Ditz suffered from a credibility gap during the campaign. Although both laid bare the Achilles heel of the government's economic policy, neither could escape the charge that they had been co-responsible in its implementation with the SPÖ. On television Schüssel looked modern and sounded eloquent but the hoped-for break through in the polls failed to come. Unlike Vranitzky, Schüssel reserved the option of doing business with the Freedom Party after the election. Vranitzky in the end, it seemed, was the most convincing candidate for the anti-Haider camp.

The Freedom Party campaign centred exclusively around Jörg Haider. He followed a gruelling campaign trail covering the length and breadth of Austria, often returning late at night to the studio in Vienna for vital television confrontations. Of particular interest to voters was whether Haider would form a government possibly with the People's Party. The FPÖ veered from going for a policy of outright opposition after the election to considering possible power-sharing. Observers remarked that the lack of a clear strategy possibly confused voters.

Haider rammed home the themes which had brought him success in the 1994 election promising to "clear out the muck" in Austrian politics. He cited alleged cases of privilege, nepotism and "putrid corruption" in public life and especially the federal reserve bank, using his by now familiar flash cards technique on television. During debates Haider would pull out large cards visually depicting his points in block capitals for viewers. This ruse caught on, becoming almost a national sport; it was aped by politicians from different parties. One problem was that this had now become all too familiar for a public hooked on political "infotainment." Despite late night screening television duels drew a record audience.

However, much like a fickle Wimbledon tennis public, new stars and stunts were in demand each year by a critical electorate. Many of Haider's allegations were challenged the following day leading to a flood of law suits and bitter recriminations. In the end it was difficult to see which party was selling the best "true lies." The style of the campaign became increasingly populist and "Haiderised." A leading People's Party member called the 45 year old Haider an "aging playboy with fascistic thoughts." Demagogy and class warfare rhetoric punched its way onto the Austrian television screens. Conservatives talked of "bashing the reds" and the left did its fair share of tub thumping with cries of "clobbering the rich." People's Party politicians took to the streets to support a farmers' demonstration and railed against the socialist "golf course" chancellor.

For the FPÖ the election was premature in many ways. The party was fixed on a "1998" project for a Third Republic geared to getting into office and implementing a Gingrich-type of "Contract with Austria." This envisaged more direct democracy, less government, low taxes, less subsidies and a better deal on Austria's budgetary contributions to Brussels. One problem for Haider was that the issues which dominated the campaign such as the economic situation and the federal budget were not those which he could most effectively exploit. Topics such as scandals and corruption affairs or immigration were rated less important by voters than social justice and secure pension rights.

This election caught the party off balance with many ideas for the transformation of Austria only half-developed. The "Contract with Austria" project was scheduled for launching in early 1996 and was in embryonic stage. It suddenly had to be hastily assembled and presented to a public more interested in the financial cut backs proposed by the coalition than with intellectual sophistries with a trans-Atlantic parentage. The Freedom Party election platform consisted of "*20 pledges for the Contract with Austria.*" These policies were to be put through in the first phase of any possible government after the election. However ideas for the renewal of Austria suddenly sounded too radical for an electorate looking for reassurance and security. The time for change theme scared many voters fearful that in the turmoil they could lose out. The safety first strategy of Vranitzky played to the basic conservative instincts of Austrians or as *The Economist* put it, "Vranitzky, a former bank official, oozes decent stolidity, consensus-minded complacency and fear of radical reform."

Another handicap for Haider was that he had become a victim of his own success. The intoxicating gains of the 1994 election had hardly been absorbed and many in the party were still in a state of euphoria. The party was enjoying its success, making the most of new instruments of power available to it such as calling special sittings of parliament. It was constantly in the limelight, a thorn in the coalition's side and rated by the public as a useful check on the government. It was questionable however whether most wanted to see Haider and his ensemble take over and run the country. People were told this would ruin the country's image abroad painstakingly repaired after the Waldheim affair, and were also uneasy about what Haider actually would do if he became chancellor. In an interview with *Der Spiegel* just before the election, Simon Wiesenthal stated it would be a "catastrophe for Austria" if Haider were to come to power. Wiesenthal had just received an honorary citizenship of Vienna and was a respected figure who had dedicated his life to tracking down Nazis. He was convinced that Haider was always looking for something positive in National Socialism:

> Haider's parents were Nazis through and through. Much of what he says that is so uncontrolled, he heard as a child at home. His parents certainly didn't talk of concentration camps, but of "penal camps." His party is a *Führer* party and he is a dictator in a democratic guise.

The Freedom Party in government, according to Wiesenthal, would wreck the Austrian tourist industry (see the interview with Wiesenthal in this volume).

Haider's reputation was based on constant movement which some considered a destabilising form of negative energy. By contrast Vranitzky looked a safe pair of hands who might not inspire but could at least be relied on to administer. In reality the stereotypes were misleading. The early election was not the result of Freedom Party pressure but the internal contradictions of the coalition. Vranitzky's government had failed to manage effectively the economy and had contributed to the alarming budget deficit. Now the same people who had largely been responsible for the shortcomings of the past suddenly came forward to rescue the country in its hour of need. With some ingenuity the government managed to convince voters that the real danger to stability was not the parlous budgetary situation but Jörg Haider. The opponents of Haider had either scared themselves into believing that Haider was unstoppable or deliberately built him up as public enemy number one. In the past Haider had flourished on this negative publicity and could play the wounded underdog who was unfairly treated by envious rivals. In 1995 Haider was no longer the innocent onlooker but was in a position to make a real challenge for power and shake up the post-war Austrian system.

This phantom picture of the Freedom leader overlooked the internal problems of party organisation and weaknesses in strategy which were to surface during the campaign. Early election posters showed Haider under a biblical style caption of "HE has not lied to you." Unfortunately as many pointed out the "Simply honest, simply Jörg" was unaccountably looking away from the camera and not straight into the eyes of his public. Haider himself it seemed was not really sure if the time was ripe to go for power or whether it was best to develop the policy of a strong opposition. As a result there was a failure to develop a clear line on the position and role of the party after the election.

At the beginning of the campaign in an opening rally in Klagenfurt, Haider staked out a clear bid for leadership before a jubilant crowd of five thousand fans. Many who saw clips on TV said it reminded them of Nuremberg, an exaggerated comparison which had nonetheless become typical. Haider made some observers nervous with his call to "clean out the muck" in government and high places and establish "order." "I am ready" he continued, " to take on the leadership (*Führung*) of this country." As *The Financial Times* commented, "Somehow when he says *Führung* it sounds different from when someone else says it and he knows that it sounds different."[3] Although subsequently Haider toned down the radical rhetoric of his Klagenfurt speech, the nervousness remained. This was played upon with devastating effect by Heide Schmidt, former colleague and now merciless foe in a television debate live with Haider. As a way of countering Haider's flash cards which in any case were now wearing thin and showing a tendency to fall over, Schmidt delved into her handbag pulling out a concoction of recording equipment and snarled up cables. More by chance than judgment it seemed, she pressed the correct button before a decidedly uncomfortable looking Haider and intrigued interviewer. Several minutes followed in which the eerie sounds of Haider's voice crackled from the tapes. The Klagenfurt speech with its references to leadership and cleaning out the muck were followed by frenetic applause. Schmidt had used Haider's own words to support her own case that he was dangerous, sinister and ruthless.

The reference to muck-cleaning was to become the leitmotiv of the Freedom campaign. It was made to sound scary but again the problem was probably Haider rather than the message. At the beginning of 1996, Paddy Ashdown for his Liberals called for a similar cleaning up campaign in British politics to get rid of the muck and corruption. No-one found anything particularly sinister about this sentiment. During the campaign, Haider received many gifts of manure forks and buckets from an admiring band of followers to help him in his project. Many sympathised with his aim to cut back on abuse and privilege at the top when ordinary people were having to tighten their belts. This message could be easily grasped. It was the outright bid for the chancellorship which caused the party most problems and attempts to change the policy ended in further confusion. Eventually Haider seemed to favour the policy of a strong opposition which in effect remote-controlled the government of the day. It was an effort to please those who did not want to see the party permanently consigned to opposition and reassure those who felt that the time was not ripe for government.

Haider was not as invincible as his opponents feared or made out. The election night was an obvious disappointment to a party unaccustomed to losses however modest. Haider's winning streak had been halted and the image of eternal youth was beginning to crack under the strain. This was a new experience for Haider used to being portrayed in the press as athletic, sexy and handsome. *The Spectator* of 21 October described him as "Europe's most successful far-right politician" and as a "pin-up populist. Viennese magazines frequently run photos of him bare-chested or preening in the briefest of bathing trunks. Paparazzi chase after him scaling rocky mountain faces, white-water rafting or engaging in a host of other swashbuckling pursuits." The author continued to relate how Haider turned up for an interview "wearing skin-tight shorts patterned in hot purple, red and orange." *The Sunday Times* even went so far as to run a story under the heading of "Austria's neo-Nazi sex symbol pulls floating vote." Haider, the reader was informed, "is packaging himself as a sex symbol. He prefers denim shirts and jeans to the suits worn by other Austrian politicians" – this comment appeared next to a picture of Haider wearing a thoroughly respectable dark suit and sober tie.[4] The article revealed that "there are only one or two occasions a year when he lets his wife appear with him. Normally she is tucked away in a corner so nothing comes between him and the women in the crowd." Myths such as these about Haider abounded despite the fact that most of Haider's fan club and those who showed up at rallies were predominantly male and on the wrong side of forty.

Increasingly Haider was being confronted with politicians from his own generation. He faced new men from the Social Democrats like Viktor Klima who stood in for Vranitzky in a television debate during the campaign. Klima although slightly older than Haider looked fresh and youthful and could display the same kind of dynamism, good looks and pugnacity. Early in 1997 Klima took over as chancellor in place of a weary Vranitzky.

THE RESULTS

The main loser of the election was the Green Party under Madeleine Petrovic who had been a popular star of the 1994 campaign. She appeared stilted and lifeless in many of the debates and distinctive "green" issues were submerged by an obsession to stop Haider and the formation of a possible "bourgeois bloc" with the People's Party. Those frightened by this prospect however tended to vote for the most credible anti-Haider party which was perceived to be the SPÖ rather than the Greens. After the election the Green project went back to the drawing board with internal squabbling between the "realos" and the "fundis." The Greens dropped to become the smallest party represented in parliament and were overtaken by the Liberals under Heide Schmidt. They only just managed to clear the 4 percent hurdle necessary to qualify for seats in parliament.

Despite overtaking the Greens, the Liberals too fell far short of their electoral goal. They campaigned as the "offensive centre" in Austrian politics championing the rights of women and homosexuals. Many of their policies were identical to those of the Greens with the exception of some liberal economics. They had played the role in the outgoing parliament of the loyal opposition and had gone along with the coalition's pro-European line before the referendum. After the poor election result, Schmidt came under fire from her own members for her "authoritarian" leadership style. One problem is that like Schmidt many Liberals have a past as former cohorts of the FPÖ. Whilst their position is now far removed from the Freedomites, the Liberals find it embarrassing to be reminded of their old allegiances.

The only real winners of the 1995 election were a jubilant Franz Vranitzky and his Social Democrats. Schüssel made modest gains but the result fell far short of putting him in the federal chancellory as the "number one" (see Table 1). Most of the socialist gains came from the mobilisation of previous non-voters. The electorate was seduced by SPÖ promises of looking after old age pensions and the less well-off but the result left open the perennial question of how this could be financed. Many asked after the election whether it had all been worth the fuss. Austria was left eight weeks later, after a divisive campaign, with no obvious government and no quick fixes for the budget deficit. The Social Democrats after 25 years in government looked just as cocky as ever, the People's Party's casino politics had backfired and no-one really believed that Haider had been stopped. Those who swung back to the SPÖ in 1995 forsook them at the next opportunity.

For many the Freedom Party had become the "underdog party" and has a legitimate claim to speak for the workers and those on low incomes. People voted for the FPÖ firstly because it exposed scandals, secondly because it stood for savings and was against abuse of social welfare, thirdly because of its position on immigration and then because of the personality of Haider.[5] According to an analysis by the political scientist Fritz Plasser, 34 per cent of workers turned out in 1995 for the Freedom Party (see Tables 8 and 9). This trend continued in the Vienna elections and the elections to the European Parliament in October 1996. Plasser concluded that the FPÖ had become a "protest oriented workers' party of a new type." Overall the Freedom Party dropped slightly in percentage points but gained more votes than in 1994. In the provinces of Salzburg, Vorarlberg and Tyrol it actually gained in percentage terms but its performance nationally was brought down by losses in eastern Austria. Here the socialist dependency network had greater roots and could be expected to operate during a time of *Angst*.

In 1994 the traditional two party system had been usurped by a pentagonal structure.[6] The 1995 results saw a re-stabilisation of the two old parties, the SPÖ and the ÖVP, but a consolidation of most of the gains of the Freedom Party.

NOTES

1. *The Financial Times*, 12 December, 1995.
2. *The Economist*, 9 December, 1995.
3. *The Financial Times*, 11 November, 1995.
4. *The Sunday Times*, 17 December, 1995.
5. Information from F. Plasser, P. Ulram, E. Neuwirth, F. Sommer, "Analyse der Nationalratswahl 1995," in "Politische Macht und Kontrolle," nr. 10, 1995-6, *Informationen zur Politischen Bildung* (Vienna: Jugend & Volk, 1996), pp. 109-120. Plasser believes that the personality of Haider cannot be disentangled from these other motives which have been penetrated to a greater or lesser degree by the "Haider factor" (interview). See also F. Plasser, P. Ulram, G. Ogris, *Wahlkampf und Wählerentscheidung*, (Vienna: Signum, 1996).
6. See M. Welan, "Regierungsbildung," February 1994, discussion paper nr. 26 of the BOKU University, Vienna.

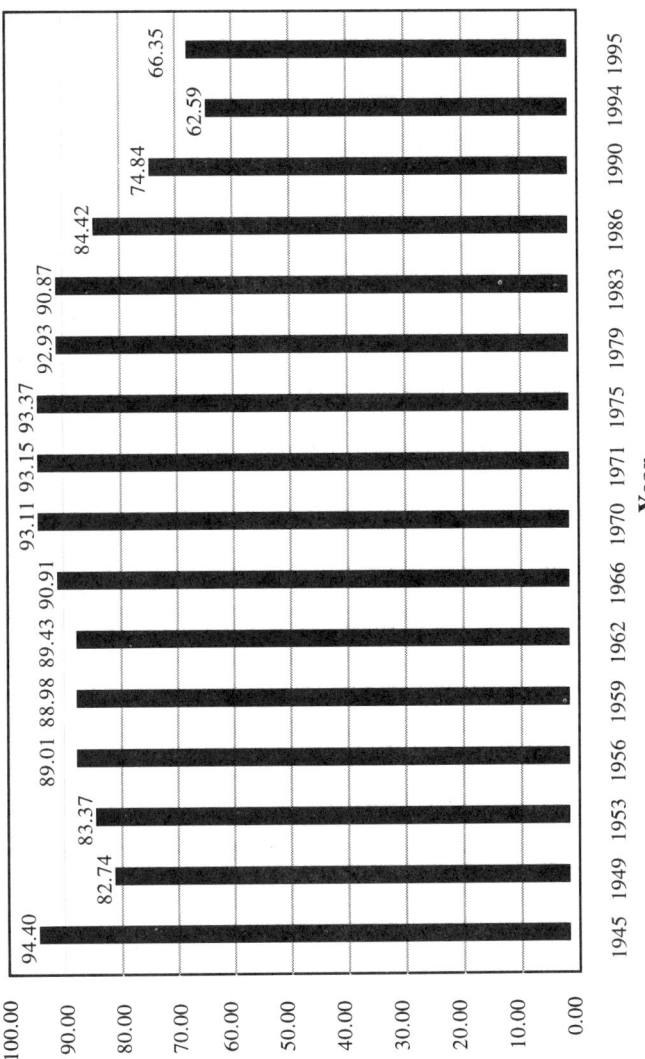

FIGURE 5
COMBINED SHARE OF VOTE, ÖVP AND SPÖ, 1945-1995

Source: Austrian Statistical Office.

TABLE 8

SOCIO-DEMOGRAPHIC BREAKDOWN OF VOTES FOR THE FPÖ, 1986-1995

% Voting for the FPÖ in the Election for the Nationalrat	1986	1990	1994	1995
Men	12	20	28	27
In work	13	20	28	30
Pensioners	11	22	29	23
Women	7	12	17	16
In work	7	13	17	20
Housewives	8	11	17	14
Pensioners	5	12	19	10
Aged under 30	12	18	25	29
between 30 and 44	11	15	22	24
between 45 and 59	6	15	22	10
over 60	8	16	22	15
Farmers	5	9	15	18
Self employed/professions	15	21	30	28
Civil servants, public service	9	14	14	17
Salaried employees	13	16	22	22
Blue-collar workers	10	21	29	34
Pensioners	8	16	24	16

Source: F. Plasser, P. Ulram, G. Ogris (eds), *Wahlkampf und Wählerentscheidung*. Signum, Vienna, 1996, p. 175.

TABLE 9

CHANGES IN ELECTORAL BEHAVIOUR IN NATIONAL COUNCIL ELECTIONS, 1986-1995

Group	Year	SPÖ	ÖVP	FPÖ	Greens	Liberals
White-collar workers	1986	40	36	13	7	-
	1990	38	27	16	7	-
	1994	29	25	22	12	11
	1995	32	28	22	7	8
	Change, 1986-1995	-8	-8	+9	0	(-3)*
Blue-collar workers	1986	57	26	10	4	-
	1990	52	21	21	2	-
	1994	47	15	29	4	2
	1995	41	13	34	3	4
	Change, 1986-1995	-16	-13	+24	-1	(+2)*

* Change 1994-1995.

Source: Plasser, F., Ulram, P., Neuwirth, E., Sommer, F. "Analyse der Nationalratswahl 1995" in Politische Macht und Kontrolle, nr 10, 1995-6, *Information zur Politischen Bildung*, Jugend und Volk, Vienna, 1996.

20 Pledges for the "Contract with Austria"

Contract with Austria

Austrians have been lied to in the past by the ruling coalition Socialist and People's Parties (SPÖ/ÖVP). They have broken all their pre-election promises before: on tackling the alarming budget deficit, on pensions' reform, on administrative savings, on health reform, and on modernising the railways, etc., etc.

We in the FPÖ (Austrian Freedom Party) have not duped the electorate and pledge in the future to keep our word in a signed "Contract with Austria."

This should be put forward in the first hundred days of any new parliament.

Our initiative consists of the following points:

1. A Pledge to Thrift

The constitution should be amended to limit the federal debt to 60% of the Gross Domestic Product (GDP). The maximum takings or "annual state grab" (taxes, fees, duties, etc.) should be limited to 40 % of the GDP instead of the current 44%. This amendment should be put to the people in a referendum. A transition period of three years is foreseen for these constitutional upper limits to become operational. Any government then violating these norms can be dismissed by the federal president.

2. A Pledge for more Democracy

A petition drive supported by over 100,000 citizens should suffice for a mandatory referendum, as in Switzerland.

At present the forms of direct democracy such as the petition drive or referenda are ultimately controlled by the governing parties rather than the people. Parliament is only required to discuss, but not act upon, a successful petition drive and a referendum needs the consent of the government.

Not only the federal president, but also key politicians at state and provincial levels such as governors and mayors, should be directly elected by the people.

3. A Pledge for More Freedom
The question of abolishing compulsory membership in the chambers should be put to the country in a referendum.

At present people are forced to join the chambers and various other professional institutions. We believe this is archaic and infringes the right to freedom of assembly.

4. A Pledge to Safeguard Freedom of Thought
Austria operates the last state-run public opinion monopoly in Europe. This has already been condemned by the European Court for Human Rights. Private radio and television stations should be allowed.

5. A Pledge for More Justice
The current practice whereby the Executive can direct or terminate legal proceedings shall be abolished. The appointment of High Court judges should not be influenced by political parties.

6. A Pledge for More "Checks and Balances"
Adverse findings and criticisms of the Central Auditing Authority must lead to action. This body is currently limited by parliament. This is a "watchdog" which can bark but not bite.

7. A Pledge to Prune Government Administration
Austria leads the world in public administration and costs. Many services can be contracted out to the private sector.

New laws and regulations should be cost-accounted. Reductions in the public payroll can be achieved through a hiring freeze. This forms part of a long-term consistent savings programme.

8. A Pledge for Lean Government
Bureaucracy has a habit of proliferating. Austria has three times as many ministers as Switzerland. Rationalisation in government could lead to more efficiency, and savings in luxuries such as ministerial cars and perks.

9. A Pledge to Slash Privileges

Politicians' privileges including exorbitant golden handshakes, severance pay and multi-pensions must be abolished. Here government politicians get more than the US President and do less. Leading politicians in our party have voluntarily limited their salaries.

"Jobs for the boys" should be abolished and appointments in the Federal Reserve Bank and the state sector should be made on the basis of merit rather than the possession of the right party book.

10. A Pledge to Cut Subsidies

Austria is subsidised to death. More state and more taxes have solved nothing in the past. Many subsidies go to political hacks. The public funding of political parties should be halved and state support for the press should be totally eliminated. The latter is currently being examined by the European Union on the grounds that it distorts free competition.

11. A Pledge to Cut Taxes

Austria is overtaxed and over-regulated. Both damage the competitiveness of the Austrian economy and restrict the purchasing power of consumers.

The tax system should be simplified and taxes cut. The top rate should be cut by three percentage points and for lower salaries by five points.

12. A Pledge to Safeguard Economic Competitiveness

Austrian industry suffers from high supplementary wage costs, permanent reductions in investment incentives and tax barriers which hinder the growth of domestic capital resources. The sell out of core areas of Austrian industry is well advanced. This has led to job losses.

An active economic policy must foster private economic activities. Tax incentives must be selective, promote research and development and be accompanied by administrative reforms.

13. A Pledge to Safeguard All Earned Pension Benefits

Austria holds the world record in early retirement schemes. The average retirement age is around 57 and going down. The system is widely abused and many so-called pensioners are in good health and continue to work doing lucrative part-time jobs in addition to pulling in the state pension. This waste and abuse has to end.

A pension reform should have at its core the allowance of private provisions for retirement on top of those provided by the state. The transition to such a system has to assure the retention of full benefits already accumulated and paid for by employees to date. For pension benefits already earned, there shall be no reductions or other "penalty" payments.

Older employees should be protected from early layoffs by a combination of partial pensions and continuance of salaried income from their firm.

This would be an alternative to the current practice of early retirement.

14. A Pledge to Alleviate the Tax Burden on Families

The current system is unjust. A single monthly income is taxed much higher than the same income accrued from two jobs. We propose a minimum "per capita" tax free family income code. This would eliminate the current discrimination against those families who are less well off.

This reform should be achieved through savings in the cost of government and not through greater tax burdens on singles.

15. A Pledge to Safeguard the Health Service

The system is currently prone to waste and money is allocated not on the basis of performance but without any relation to services rendered.

The actual budgeted cost for surgery or health services shall be reimbursed instead of the current practice where hospital stays are automatically reimbursed, however long it takes. Similarly a services based reimbursement of private doctors would take the pressure of the state hospitals.

16. A PLEDGE TO PRIORITISE THE SECURITY OF AUSTRIANS

Crime is on the rise but rates of conviction are sinking. Organised international crime is rising explosively. The police must be equipped with the same modern technology as the criminals. Sentences have to be harsher to have a deterrent effect and stop criminals from committing the same act again. For capital crimes, life long sentences should mean life sentence with no early release for "good behaviour." Foreigners sentenced for criminal acts should be deported.

17. A PLEDGE TO LIMIT IMMIGRATION

The existing immigration laws should not be softened up. Each potential immigrant should give proof that they have a job and accommodation. To enable the return of the numerous illegal immigrants in Austria, we propose ID or identification requirements such as exist in most industrial countries today.

The current practice of granting citizenship well before the legally required ten year waiting period should be discontinued.

18. A PLEDGE TO STEM THE LOSS OF RIGHTS TO BRUSSELS

The inter-governmental reform conference next year will determine whether more powers should be surrendered to the European Union or whether Europe should become a confederation of states in which the member countries retain sovereignty. Austria should fight for the continuance of the unanimity principle and limit the flow of power to the Brussels bureaucracy. Austria's vital interests must not be overruled by the EU.

Austria should seek, as the United Kingdom did, to re-negotiate its high budgetary contribution to the EU coffers with the aim of reducing these payments by one third.

Efforts to enforce clauses in the Maastricht Treaty which would mean majority rule by the EU over Austrian water resources must be resisted.

19. A PLEDGE TO ALLOW TENANTS OWNERSHIP OF PUBLIC HOUSING

Such tenants fork out the same as if they owned the dwelling outright. We would change the housing laws to allow tenants to become the rightful owners once they have made these payments, since there exists no reason to deprive them of such ownership.

20. A Pledge to Safeguard the Environment

Environmental laws shall be reformed along the principle of "the polluter pays." When pollution emissions become expensive they will be reduced.

Environmental burdens imposed on agriculture shall be reimbursed to farmers directly. This would protect farm incomes and at the same time reduce the use of environmentally damaging substances.

10

TO KRUMPENDORF AND BACK

The 1995 election was noteworthy because it showed Haider's path to the Vienna chancellory could be stopped. A postscript to the election soon became associated with the Carinthian town of Krumpendorf and showed once again the pitfalls that opened for Haider in dealing with Austria's brown past.

Just before election day in 1995, a controversial amateur video was broadcast on German TV showing Haider addressing war veterans with former members of the Waffen SS in the town of Krumpendorf. They had congregated from Belgium, Norway, Denmark, France and Germany in the spa town every year for a reunion. The video had been made in September 1995 during an official traditional meeting in Carinthia.

In the video Haider appeared to warmly embrace the audience as "dear friends" and spoke of the decency, intellectual superiority and honesty of those who were not yet in a majority. Austrian television declined to show the clip until after the election to allow for it to be properly authenticated. The FPÖ complained that it had been deliberately doctored to discredit the party and wreck a possible coalition with the People's Party. The recording showed photos of former SS members who had not been present and included a superimposed commentary which gave the impression that the audience consisted of old Nazis. Haider later claimed that he had not been praising Hitler's SS (as was widely reported) but his own party workers.

In an initial reaction, Haider caused a further furor in a television interview for standing by the war generation whom he argued could not be criminalised en bloc. He enraged his critics even further by describing the Waffen SS as part of the Wehrmacht (the regular German army in the Second World War) and concluded that a speech before such a gathering was in no way questionable. His comments were seen as a whitewash of the Nazi period, although Haider insisted he was just refuting the idea of collective guilt. Simon Wiesenthal along with others criticised Haider's lack of historical knowledge and declared that the Waffen SS had been a murder squad. The Documentation Archive of the Austrian Resistance in Vienna noted that the Waffen SS was declared at Nuremberg to be a criminal organisation.

The FPÖ replied that according to the Ministry for Social Administration in 1946, "the Waffen SS was not a military unit [*Wehrverband*] of the NSDAP but a part of the Wehrmacht." Service in the Waffen SS was recognised for Austrian pension insurance purposes whereas duties in the SS were not. Haider's position was that he had not praised the Waffen SS but had opposed general outright condemnation of those who had served in those difficult times. According to Haider, there could be nothing wrong in showing respect for the older generation who served in the war.[1]

The Krumpendorf incident opened old wounds making Haider vulnerable to charges that he was deep down a Nazi with only a thin democratic gloss. This came at the end of a year when he had been carefully cultivating relations abroad aimed at portraying a respectable, politically correct, and democratic image. Now these efforts seemed to have been in vain: the London *Times* of 18 December 1995 ran an article under the heading of "Austrian electors stifle ambitions of pro-Nazi leader."

The Krumpendorf speech and the election results of 1995 plunged the party into a fit of gloom and introspection. The bandwagon had ceased to roll and many projects were quietly buried or put on the back burner. Initially it was rumoured that Haider might have to resign or face court proceedings for his comments. In 1992 parliament had passed an amendment of the constitutional law of 8 May 1945 which prohibits the NSDAP (National Socialist German Workers' Party). The key passage of the amendment stipulates penalties for "persons who contest, seek to diminish the importance of, endorse or attempt to justify the genocide carried out by the National Socialists or other National Socialist crimes against humanity in the print, electronic or other media or in any other way calculated to reach a wider public." Up to ten years imprisonment was possible for those found guilty under this clause. The Greens pressed for legal proceedings against Haider on these grounds and some Social Democrats clamoured for his resignation as a member of parliament. After the initial fuss nothing substantive could be found to force this action.

An uncertain future now faced the FPÖ and the general despondency was reinforced by the news that the party had forfeited the right to election funding to the tune of 30 million Schillings because of a failure to submit its application on time. In parliament the party had lost two seats, dropping to 40 and falling automatically into a lower bracket for financial support. The party was more hopelessly isolated than ever before and its chief withdrew for a long Christmas vacation.

This was not the first setback for Haider and the party. There had been a similar ebb in morale after the loss of the provincial governorship and after the "Austria First" popular initiative. Each time Haider had bounced back with sensational electoral gains and renewed confidence. A leading psychoanalyst, Erwin Ringel, had once attributed Haider's success to his anti-depressive effect. According to this theory, Haider had undeniably pulled Austrian politics out of the depths of boredom. Psychological experts like Ringel maintained that boredom is a form, or concealed form, of depression. Haider, the entertainer, works as an anti-depressant on the Austrian psyche with its innate predisposition to depression and suicidal tendencies. Now it seemed the FPÖ boss himself was in need of some special Haider medicine to lift the winter blues. The newspaper *Kurier* remarked, under the title of "Jörg Haider's Winter Depression," that the boyish grin had disappeared and the voice had become quieter and less ebullient.[2] The party was not involved in the formation of a new government and seemed no longer to be pulling the strings in domestic politics.

Many followers of Haider privately criticised the Krumpendorf appearance, unable to understand why he had been so clumsy as to be filmed in such company. Leading members of the party were unsure how to react officially to their leader's faux pas. Differing interpretations of their beleaguered chief's comments were offered and sustained with the general conclusion that it was a deliberate conspiracy to undermine the credibility of Jörg Haider personally. It took some time before an agreed line emerged which looked at all convincing. The leader of the FPÖ in Vorarlberg, Hubert Gorbach, stated that he was against all generalisations and just as there could be no collective guilt there could be no collective innocence either. He rejected all unqualified attempts to glorify the SS and any efforts to excuse what had been done in its name. Under pressure from the ÖVP whose goodwill he needed in the province, Gorbach was one of the first to try a clean explanation. He categorically recorded that the FPÖ condemns all attempts to play down the incidents which took place under the National Socialist regime and rejects any promotion of this thought. His position was acceptable to the ÖVP federal leader, Wolfgang Schüssel, who was hoping the FPÖ would not put itself totally out of bounds for the rest of the legislative session.

Haider's initial response had done little to help the party out of its Krumpendorf dilemma. President Klestil intervened in the affair and called on Haider to react with sensitivity to criticisms of his remarks concerning the role of the Waffen SS under the Nazi regime. In unusually sharp tones the federal president went so far as to say that these comments went against the grain of the democratic basic consensus of the Second Republic which understood itself as the antithesis of every type of totalitarian rule. He called upon Haider to find the right words "in the interests of the republic and in his own interest." Haider replied that he was only concerned that the entire war generation should not be as a whole condemned, a generation which had after the war found its way back to democracy. For Haider the matter had nothing to do with the role of a particular unit in National Socialism since there could be no justification for that system or its separate components.[3]

In a press conference on 9 January 1996, Haider reiterated that he had never praised the Waffen SS and showed some irritation at the entire debate which he believed was designed to thwart a possible coalition with the ÖVP. He maintained that, "I have always fought against dismissing the war generation as a whole as criminals." Again in a party New Year speech in Linz in 1996, Haider stated he would resist all attempts to sow the seeds of hate between generations.

Haider pointed out that he had repeatedly condemned the horrors and atrocities of the Third Reich in no uncertain terms in speeches and in his book, *The Freedom I Mean*, where he wrote:

> The most horrific aspect of National Socialism was its anti-Semitism, racism and mass extermination programme. These crimes cannot be made good. Germany tried in the form of material compensation. Austria with its "victim" theory passed the buck. Only fifty years after the war did an Austrian chancellor, the Socialist Franz Vranitzky admit that Austrians were involved in those crimes. I reject collective guilt, I accept joint responsibility.

Haider continued:

> Nothing can ever justify National Socialism. There can be explanations for its rise and success.... For this darkest chapter in their history, the Austrians can bear no collective

guilt but they do have as much responsibility as the Germans. We cannot close our eyes to seven terrible years and their legacy. To do so means erasing our identity, the positive as well as the dark sides.

In a speech in the Spring of 1992, I outlined my basic political standpoint. I said: 'Whoever goes with me stands for an Austrian Freedom Party without brown stains, but also without fear of historical truth. Whoever goes with me stands for an Austrian Freedom Party with credible distance from the time of National Socialism but with respect for the older generation, which found the way to democracy after their own bitter experiences.'[4]

At the beginning of 1996 some effort was made to repair the damage and a party political broadcast was screened on television:

> In order to distract attention from the real problems of our country, the SPÖ and ÖVP try to discredit the Freedom Party, Jörg Haider and over a million voters.
> The appearance of the party leader at an official event as part of the Ulrichsberg meeting has been criminalised with the aid of a manipulated TV report.
> In an interview which appeared on 14 December 1995 in the German magazine *Bunte*, the FPÖ leader made his position on National Socialism clear, as so often, without any ambiguity:

I always describe the Third Reich as the most heinous criminal regime. The worst thing about it was anti-Semitism and racism which led to mass extermination. Such crimes cannot be made up for. National Socialism cannot be justified by anything.

> On 6 January 1996, the provincial leaders of the Freedom Party met in Linz to underline in a joint declaration the comments of Dr. Haider:

We condemn the attempts to exclude the Freedom Party and their representatives from the democratic basic consensus by means of defamation, manipulation and deliberate misinterpretation.
It is completely indisputable that the Freedomites clearly condemn all attempts to play down the events which took place under the National Socialist regime and that they utterly reject every totalitarian and undemocratic thinking.

> *The Freedom Party is aware of its duty and responsibility to the older generation, whether it be to the persecuted, the displaced persons or to the participants in the war and we reject collective guilt as much as collective innocence.*
>
> *We will not allow this generation or a part of it, which through no fault of its own had to overcome the most difficult times of this century, to be condemned en bloc.*
>
> *Nothing else than this was expressed by Dr. Jörg Haider in his explanations. Therefore we strenuously reject all efforts to twist around the statements of the Freedom leader to mean the opposite.*

— end of broadcast —

In February 1996 the party finally started to go on the offensive. It produced documentation to prove that Nazi skeletons lurked in the cupboards of the other main parties and particularly went for the Social Democrats. A pamphlet was produced (*Protocol of Calumny*) which carefully reviewed the background to Krumpendorf and Austria's past, including the role of the governing establishment parties. Whilst such a tactic could not condone any party but simply convey to the outside observer that all were probably as bad as one another, this ploy had the desired effect of putting the Socialists in particular on the defensive and defusing a potentially dangerous situation. It is doubtful whether outside of Austria this debate had much effect since by early 1996 the fallout from Krumpendorf was widespread.

The argument that the two big parties had taken a lenient stance on the country's role in the Holocaust, and during the Nazi period, was common knowledge. The distinguished political scientist Anton Pelinka noted in his book *Windstille* that chancellor Kreisky had taken into his first cabinet ex-members of the Nazi party.[5] These included his Agricultural Minister, Johann Öllinger, who had not only been a member of the NSDAP but also in the SS from the beginning of the 1930s. Pelinka acknowledges that Kreisky had not been aware of Öllinger's SS past, but still publicly defended him until he resigned ostensibly for health reasons. His place as Agricultural Minister was taken by Oskar Weihs, also a former Nazi party member. The first Kreisky cabinet intake also included former NSDAP member Otto Rösch who became Interior Minister. He escaped detection as a former Nazi and started a political career with the SPÖ. Rösch was not only Interior Minister but in 1977 became

Kreisky's Defence Minister despite his past. Haider often liked to be thought of as the natural heir to the Kreisky tradition, although it was doubtful if this is what he had in mind.

The ÖVP too had often turned a blind eye to the brown past. Pelinka recalls how the former ÖVP mayor of Mayrhofen, Franz Hausberger, in a tourist resort in the popular Zillertal of Tyrol had been known as a volunteer in the SS from 1941 to 1943 and active in Russia and Holland. Amongst his duties was the administration of concentration camps in Holland. Mayor Hausberger attended the meeting of the Tyrolean tourist board in Mayrhofen in 1984 in which a Dutch journalist was insulted by him with the words "this Jew Journalist should get out." As Pelinka notes, "no Tyrolean politician, no Austrian politician found it necessary to stand up in public against the mayor of Mayrhofen."[6]

Friedrich Peter, former SS member and subsequent leader of the FPÖ, was similarly leniently regarded by Kreisky when it came to considering a strategic "coalition" between their two parties. Although this did generate some controversy, the point was that the SPÖ had opened the door in principle to deals with the Third camp, even with those who had been implicated in the Holocaust. The unit to which Peter had belonged had the job of liquidating civilians, especially Jews, at the eastern front, although it could never be shown that he personally took part in mass murder. It was of course possible to argue that people like Peter had learnt from the past and had become reformed liberals and democrats. However, Pelinka in his analysis concluded that both the established parties had been manifestly negligent in dealing with the past: "The problem of this republic is the conscious sloppy manner in which its representatives have dealt with everything which is opposed to the democratic republic."

This sentiment was echoed by the FPÖ brochure *Protocol of Calumny* which tried to explain away the Krumpendorf affair which was weighing the party down like a dead albatross. According to his booklet, the venue in Krumpendorf had been organised by the "Kameradschaft IV" (KIV) which mostly drew its members from former members of the Waffen SS. The KIV had been allowed to form in 1953 and after 1979 was open to all soldiers. The FPÖ refuted the charge that this was a secret meeting of SS veterans as had been reported in the German and British media. The relevant authorities had been informed in advance as was customary, the participants crossed the border openly and representatives from other parties had been invited. Notice of the event had even been posted up at the local railway station.

The Krumpendorf reunion was part of the veterans' Ulrichsberg festival – an annual gathering which was attended in 1995 by the ÖVP Minister of Defence, the ÖVP governor of Carinthia plus other functionaries from both the ÖVP and SPÖ. Haider had on occasion been a speaker at this event which he insisted was intended as a time of reconciliation (it was open also to Allied war veterans). In 1985 and 1990 he had addressed the assembly and paid tribute to the soldiers of the Second World War including the Waffen SS for their contribution to freedom and democracy in Europe. In a subsequent interview Haider was questioned on this point and replied that the struggle of these soldiers pushed back the danger of communism. This gave Austria, he continued, the chance to find its way in the western democratic system and not fall foul of the communist empire. When the interviewer commented that without a war of aggression, the Soviets would not have got anywhere near Austria, Haider replied, "It's not for me to analyse this question since history then comes into play and the question has to be put, what the aggressive intentions were on both sides." Whatever the assessment of Haider's grasp of history, it is clear that he had a built-in mechanism which rallied to the defence of the fathers from the war generation. He was not exceptional in this and perhaps some of his appeal can best be explained by the fact that much of what he said, others thought but felt inhibited to speak out loud. The rehabilitation and passionate defence of the NS fathers was deeply ingrained in many of Haider's background.

Many believed that lurking genetically in Haider was an inveterate nostalgia for National Socialist ideology. His alleged unconquered past was reckoned to have stemmed from his early childhood and his parents' connections with the Third Reich. This suspicion had trailed Haider throughout his political career and had been periodically fuelled by his own lapses on the Nazis' employment policies and by the difficulties he ran into with the so-called "penal camps." The media were for the most part convinced that Haider's basic sympathies were with his "brown" family background.

In 1985 the magazine *Profil* opened an interview with him in Carinthia, "this brown corner of Austria," by asking if it was more appropriate to greet him with "Sieg Heil!" than "good morning."

"What am I supposed to reply to that?" came the response from Haider who had easily spotted the trap. Yet even as a young politician Haider was as inscrutable as a sphinx. Was he, the interviewer persisted, a genuine liberal as Kreisky had once believed or the reincarnation of a member of the Hitler Youth? By way of response Haider reaffirmed his loyalty to his father whom he argued had not only played his part as a soldier in the war but had also helped to build up the Second Republic and post-war Austria. It was this instinctive respect for the war generation which was an unshakable tenet and from this conviction at least Haider never wavered. People like his father were not "murderers but heroes" who had been prepared to sacrifice their lives for their country.

The *Profil* interview of 1985 was dominated by the "Reder-Frischenschlager" handshake. Friedhelm Frischenschlager, later founding member with Heide Schmidt of the Liberal Forum, was then Minister of Defence and a member of the FPÖ in a coalition government with the SPÖ. He personally greeted Walter Reder, a former SS officer who had been sentenced in Italy to life imprisonment, back in Austria on his release from an Italian prison, and even went so far as to shake hands with him. Reder was picked up at an airport in Graz and taken by courtesy of the Austrian military to quarters in Baden near Vienna. The republic seemed to be more generous to people such as Reder than to the victims of the Holocaust.[7]

As the chancellor Vranitzky said in June 1993 at the University of Jerusalem, "Over the years, Austria has undertaken enormous financial efforts to this end [caring for the victims of the Holocaust]; but often these efforts seemed to come out late, timid and in piecemeal fashion, as if to hide a bad conscience." He also acknowledged that those who were forced to leave their country in 1938 with the *Anschluß*, were not encouraged to return: "We can now offer a warm welcome back to Austria to all those who had been driven out of the country and whom no Austrian government in the years after the war has asked to come back. A recent change in the relevant laws, for instance, enables Nazi victims to regain Austrian citizenship without having to give up their present one and without the prerequisite of living permanently in Austria."

Frischenschlager subsequently apologised for his gesture in an Israeli newspaper. It was this apology rather than the action itself which scandalised Haider, who regarded the backdown as totally superfluous. He spoke of Reder as a "prisoner of war," a soldier like many others who had done his duty in accordance with the oath of a soldier.

Ten years later *Profil* was still trying to sort out the Haider enigma and published an article entitled, "The true Jörg – foster father of right terror or patron saint?"[8] The first interpretation was a reference to a statement from Peter Pilz, a member of the Greens in Vienna. Haider challenged the comment in the courts but lost the case and Pilz won the right to describe Haider in this way. Haider as the patron saint was an idea put forward by the FPÖ in the 1995 election in tune with the image of Haider as a kind of Robin Hood who would look after the interests of the poor at the expense of the rich and privileged. The Socialists always replied that this never squared with the real lifestyle of Haider, a Porsche-driving millionaire populist with landed estates in Carinthia. The enigma remained and the true Haider, despite journalistic inquiries, remained elusive. Haider was certainly eminently capable of reinventing himself as and when necessary. Policies could be tailor-made to suit fashionable trends and discarded if unsuccessful.

NOTES

1. Cited in *Die Presse*, 2 February 1996.
2. *Kurier*, 10 January 1996.
3. See *Die Neue Freie Zeitung*, 10 January 1996.
4. *The Freedom I Mean* (New York: Swan, 1995), pp. 42-44.
5. A. Pelinka, *Windstille* (Vienna: Medusa, 1985), p. 60.
6. *Ibid.*, p. 97.
7. See A. Pelinka, *Die Kleine Koalition* (Vienna: Böhlau, 1993), p. 46. In 1995 on the occasion of the fiftieth anniversary of the Second Republic, a special fund was established for the victims of National Socialism. In the first two years of operation payments were made to over 11,000 people, the majority of whom were resident in the USA. The foundation is administered by the Austrian parliament.
8. *Profil*, 52/3, 1995.

THE KRUMPENDORF SPEECH

Ladies and Gentlemen,
Dear Friends,

I ask for your understanding for being late but we have had a big meeting which lasted all day in Salzburg. I have therefore only just arrived here and thought perhaps that it would be good to make a public gesture, as once more Ulrichsberg and its meeting is being talked about so much; to make clear that it is no disgrace for an Austrian politician to meet the participants at Ulrichsberg; therefore I am pleased to be able to heartily greet you all once again on Carinthian soil.

Just as I came in I heard from a comrade, who said he's from Hamburg. He wanted to apologise for the nasty reception I got once at a demonstration in Hamburg, from the so-called "Hafenstraßen" delegation, or the anarchists who get rich from doing nothing and who are also being subsidised. They prepared a tough reception for me. All sorts of things were thrown and the like but I had not expected anything else. And this is also in the end the reason why I believe one must also make a counter-stance, otherwise we would really live in a world of anarchists, and that's not what in the end you fought and risked your lives for but instead so that the younger generation and youth should have a future in a community in which order, justice and decency are still principles.

I would also like to say something just because this Ulrichsberg festival is being talked about so much this year. It is always being discussed but this year especially so, because the one who should have been a guest speaker last year will come this year. Last year he couldn't land because it was too foggy. It was actually wonderful sunshine if I remember in Klagenfurt but it's always the same, one counts on short memories. Anyway a discussion arose in Austria on whether it is permissible for a minister to appear here. And that I believe is something which should make us reflect a little since we are really living in a time in which political correctness, as it is so nicely called, or moral intimidation, is being spread through the media and by those who have something to say in public life. They simply try to discriminate against those meetings and reunions of the older generation who really in the final analysis only want to reminisce jointly on what they went through together, what they experienced, and what they stand for today. And there are of course many who no longer dare to go there and join in. They say, well, maybe I could have problems if I'm there, if I'm seen. And then some politicians drift off and say I'll go there but I won't speak, or I'm not even going there, please don't write anything nasty about me. I'd just like to know whether any of these who are too cowardly to go there, or who constantly condemn the Ulrichsberg meeting, can put together a reasonable argument. There can be none, unless one gets annoyed that in this world there are simply still decent men who have character and who, even with a strong countervailing wind, stand by their convictions and have remained true to their beliefs right up until today. And that is the foundation, my dear friends, which is also being passed on to us younger ones and on which we in the end live. And therefore we have, together with you, the duty to be concerned about exhibitions like those which today come from Germany to Austria to enlighten us on the Wehrmacht. Exhibitions in which all of a sudden the Wehrmacht and members of the German Wehrmacht are portrayed as criminals. That's now common! Even in Austria there is an exhibition supported by public money – now for that we have the cash. We can spend money on terrorists, we can give money to violent newspapers, we have money for work-shy riff-raff, but we don't have any money for decent people! And so tomorrow or the day after, something will appear somewhere, since I am convinced that there are not only friends of mine in here, but also a few underground spies who will once again report what terrible speeches were bandied about. Therefore I would also like to say quite precisely and clearly: I shall always with my friends speak up so that this older generation will be treated with respect, shown respect for their lives, respect for what they have been through, and respect for all that they have preserved for us. That is something very crucial. And anyone who today chips in and says that the members of the war generation, the Wehrmacht, were all criminals, are in the end besmirching their own parents, their own family and their fathers! And a people which does not hold its forefathers in honour, is condemned to doom. But because we want to have a future, we will have to teach the lefties of political correctness that we are not to be done away with and that decency in our world still wins through, even if at the moment we are perhaps not able to form a majority. We are intellectually superior to the others and that is something very vital.

I hope that it will be a nice big meeting on the Ulrichsberg and I wish you pleasant reunions there with many friends and comrades. I assume also my parents will again be up there as it's part of a tradition. We're already abused in any case so it makes no difference. One lives anyway, as Wilhelm Busch said, "Once a reputation has been ruined, one can live on with complete abandon." I don't live so free and easy but I live better when the cards are on the table.

I hope it will be a lovely Ulrichsberg meeting for you tomorrow. I myself shall of course be up there, then I must make my way to Vienna because we have a special sitting of parliament which will take place in the afternoon and there are therefore a few things to sort out. And so I hope that tomorrow too clear words will be spoken by the festival speakers. One of the festival speakers has already said via the papers that he really would rather not speak there but his party leader has given him his orders and so he must speak. I advise him to make up his mind on the flight here. If he doesn't want to we don't need him; if he wants to then he should say what he really thinks deep down in his heart of hearts. That is a clear line!

On this note therefore a have nice evening, a pleasant stay and good luck!

TELEVISION INTERVIEW, 19 DECEMBER 1995

In response to questions concerning his appearance at Krumpendorf:

Haider: *I find it really odd that this generation is discriminated against in this way and therefore this generation has my full support.*

ORF: That means your appearance there was no mistake but you went there in the full knowledge that you were taking part in an event of the Kameradschaft IV?

Haider: *I had been invited as part of the Ulrichsberg festival just like all the other politicians. The mayors come and deputies come. The press was also represented - both the international and Austrian press. It was merely an attempt in my opinion once again before an election to develop some spectre or other against Jörg Haider. This gives me however the opportunity to make clear now that I stand by this generation and will do everything I can to see that there is no intellectual onslaught on its memorials. Since if we now start to run down this generation, then the next thing will be that while we have a memorial to the Soviet army in Vienna all the war memorials will be pulled down.*

ORF: It wasn't a completely open event, or otherwise the remark about underground spies who might have infiltrated the event would be incomprehensible. That means you expected that what you said could be made public.

Haider: *There are always underground spies, we have them at my events too. They are the ones who make a note of everything and write down who was there in order to discredit people. Since there was an official invitation and everything was well-known, I don't see why so much fuss is being made about it.*

ORF: We note you say you are happy there are still decent people who have character and who stick to their convictions - even in the face of such a countervailing wind. That's a harmless remark if said to a charity organisation but is questionable if made at an event attended by many members of the former Waffen SS. What do you say to this?

Haider: *Firstly I tell you this was an event within the framework of the Ulrichsberg organisation at which ministers, cardinals and bishops made speeches and held mass. I would also say to you that in the meantime word has got around that the audience that was present there, amongst them certainly members of the former Waffen SS, was not at all questionable since even the German federal chancellor Adenauer made it unmistakably clear in a letter meant for publication that the members of the Waffen SS were part of the Wehrmacht and therefore cannot be criticised for anything. And that with the countervailing wind – I meant it in the sense that there are these people who stand true to their love of their country and have a sense of patriotism, even if today they are confronted with publicly subsidised exhibitions which want to make them all out to be criminals.*

ORF: That means you don't find anything amiss when you say to members of an association deemed to be a criminal organisation at Nuremberg, "I find it good that you stand by your convictions?"

Haider: *Take a look at what the German Social Democrat and the SDP chairman Kurt Schumacher said on the subject of the Waffen SS and what Adenauer said and study the judgments made by the German High Courts which confirm that the Waffen SS was a part of the Wehrmacht and therefore deserves all honour and recognition which it has in public life today.*

ORF: If you refer to the '50s I have to tell you that the Waffen SS was dissolved as a criminal organisation in Nuremberg.

Haider: *I can't recall any such decision and it doesn't interest me in the slightest.*

11
F – WHAT NOW?

With Krumpendorf the prospect of an ÖVP/FPÖ coalition became less likely. During the 1995 campaign, the ÖVP had been the only party not to dismiss outright the possibility of coalition talks with the FPÖ. Haider's speech shocked many in the ÖVP who were not in any case enthusiastic about a deal with Haider. Whilst before the election, the People's Party had made much of the theme "it's time for a change" and "25 years of socialism are enough," afterwards it had little alternative but to sit round the table again with the Reds. A coalition with the FPÖ had in any case only a bare working majority and the Haider party was once again discredited. Haider now looked uncomfortably isolated and the chance of even an ad hoc partnership with an increasingly antagonistic ÖVP slipped away.

The ÖVP and SPÖ had shunned Haider before when he had overstepped the mark but political expediency often meant that permanent marginalisation was unattractive. After the provincial elections in Carinthia in March 1994 the main parties were prepared to consider dealing with Haider when convenient but then backed off. To begin with the FPÖ agreed to a People's Party governor in exchange for key administrative posts. A jubilant Haider and his acolytes had hardly begun to celebrate what some melodramatically saw as a "seizure of power" when an alarmed ÖVP cancelled the pact and turned instead to the SPÖ. This zig-zag course illustrated the dilemma for the ÖVP which wanted emancipation from the shackles of red/black coalitions. At the same time, it was afraid it would be sucked in by Haider, a character regarded as suspect and unpredictable. The ÖVP therefore left the door open to talks with the FPÖ but insisted it rejected its programmatic position as formulated in the "Third Republic."

The leader of the People's Party in parliament, Andreas Khol, had already (TV *Pressestunde*, May 1995) categorically placed the FPÖ outside the "constitutional bow" (see Figure 6), a concept borrowed from Italian politics as applied in 1946. This had placed neo-fascists and monarchists outside the bow in Italy, and originally communists although the Eurocommunists were later included. Political scientists and historians questioned whether it was possible to import the concept in this way and many were suspicious of the Khol critique.

Khol was in the vanguard of the so-called conservative wing in the ÖVP identified with law and order, family values and Catholicism. The FPÖ under Haider had dropped its previous traditional anti-clerical image and could offer a home to ÖVP sympathisers on the right. Khol's message was that the ÖVP was unquestionably a party of the Second Republic and sought as the SPÖ, Liberals and Greens to operate within the existing system. A party had to stand on the ground of the existing constitution and should seek to develop its principles to come within the bow. Khol accused Haider of ridiculing parliament and of seeking through a cultural revolution another constitution. To be outside the Khol bow was to seek an entire change of the constitution, a new political system with (legal) parliamentary and extra-parliamentary means.[1] This included the possibility of resorting to pressure of the street. Street demonstrations were rare in modern Austria and recent examples have included student and union protests against the government cut-backs and planned austerity measures. Also before the 1995 election farmers with the support of the ÖVP marched around the Vienna Ringstrasse to express dissatisfaction with the SPÖ's policy on payments. None of these were organised by the FPÖ although it had often threatened to mobilise the people in this way.

The "F movement" with its Third Republic was, for Khol, outside the bow but he hoped for the return of the prodigal party to the fold. This approach allowed the ÖVP to attack the FPÖ but held out an olive branch in the future. The FPÖ was not considered by this interpretation as undemocratic or unconstitutional, since otherwise it would be banned, but was simply outside the pale. The F sought according to Khol a "pre-fascist" Third Republic with an elected chancellor-president who would be able to dissolve parliament. It wanted a "presidential system on the US American model with a simultaneous strengthening of plebiscitary democracy." This was an ominous combination for many in Austria. Whereas the French or the Americans could be entrusted with a presidial system and the Swiss with plebiscites, many feared the outcome of such a mix in Austria. This *Angst* seemed often to be grounded in a belief that democracy in Austria was too young and had still not developed firm roots. The so-called authoritarian personality, it was thought, was dormant, the search for the "strong man" still latent and the dangers for a fragile democracy unlimited.

Khol had presented his constitutional bow on television in May 1995 and adapted the notion a year later. His analysis of the Freedom Party was backed up with references to Haider's book, *The Freedom I Mean*. This was published early in 1993 but Khol's critique only really became sharp after the FPÖ gains in the October 1994 election. The Third Republic was already established before this election as a part of F ideology but this did not dissuade 22.5 percent of the electorate from voting for the party.

Not only did Haider advocate a different republic from the Second Republic set up in 1945, he used for Khol alarmingly radical language in his book such as, "What we have in mind is more than a change of government. We want to bring about a cultural revolution by democratic means. We want to overthrow the ruling political class and the intellectual caste." For Khol this amounted to fascistic thinking. The F leader, Khol explained, was against parties and social partnership. He saw similarities with the Fascists and National Socialists, who also maintained that there was a *Führer* and the people and, in between, there was nothing.[2] Khol had also demanded from the Greens that they clearly disassociate themselves from violent acts and language of the extreme hard core left which had some links with them. The Greens obligingly did this and also had learnt within the parliamentary framework to play the game by the established rules. The Freedom Party on the other hand played all the tricks in the book in the National Council to use, and some would say abuse, the rules of procedure for its own ends.

Haider had written in his book, "The road to the Third Republic is irreversible. Post-war Austria's Second Republic is in its death throes despite the efforts of the red/black despots to protect their power bases.... The dominant political class in Austria has set up an authoritarian system which is a mixture of the Kremlin and the Vatican." Despite the reference to democratic means and a categorical rejection of the use of violence in the book, people like Khol were uneasy about Haider's real intentions. Haider had explained in the book the kind of change he had in mind:

> We want to change this system and make no secret of it. Our enemies – especially Socialists – are implying that we stand for the wanton destruction of society. Our reform programme however is constructive, designed to remedy some of the worst flaws in the Austrian party system, alien to our liberal ways of thinking. We want to get rid of the corporate

elements in this system and abolish privilege and corruption. We want to replace it with an open, democratic society of free citizens. We want to achieve this through democratic dialogue and debate.

Despite this concluding moderate tone Haider's constitutional reform programme revived fears of a presidential system based on a strong man with few parliamentary checks and balances.

The People's Party was not entirely united in its assessment of Haider and his plans for power. Just as Khol was arching his constitutional bow in 1995, a leading ÖVP politician, Josef Krainer, provincial governor of Styria, came up with a different theory. In an interview in the daily *Der Standard*, 17 May 1995, Krainer even described Haider as a reforming force in the country because of his critique of its stuffy corporatist system.

On the same day as the parliamentary election in December 1995, provincial elections were held in Styria with devastating losses for Krainer and his People's Party. He resigned as provincial governor leaving the way open for the first woman to hold this post in Austria, Waltraud Klasnic. She needed, however, the support of the FPÖ in the province to be sure of election. The ÖVP was now placed in an acutely embarrassing position with respect to the FPÖ just when the Krumpendorf speech of Haider was dominating the headlines. The Socialists were outraged by Haider's impromptu address to a dubious gathering and wanted the party placed in quarantine. Khol and his colleagues too were shaken by Krumpendorf and demanded an explanation which was duly submitted by the FPÖ in the new year declaration at Linz. Khol engaged in a detailed analysis of the FPÖ and laid down the terms for the party to be acceptable in government. The party was urged to return to the constitutional bow and the following preconditions were demanded:

- a break with the concept of the Third Republic – the basis of parliamentary democracy should not be questioned.
- acceptance of the membership in the European Union.
- social partnership should not to be called into question.
- a clear rejection of right extremism and National Socialism.
- a repudiation of the "cultural revolution," and violent diction.

Khol's constitutional bow had by now become elevated to a political model which could be periodically adapted. The FPÖ, however, interpreted this as a strategy which allowed scope for the party to negotiate with the Blacks, if appropriate in the future. The political scientist Anton Pelinka saw the "soft" concept of the bow juxtaposed to the militant, "hard" concept of the Third Republic. Pelinka cautioned, however, against taking the Freedomite model too seriously.[3] He spoke of the Third Republic as a "chimera." According to *The Chambers Twentieth Century Dictionary*, this can best be described by the following: "a fabulous fire-spouting monster, with a lion's head, a serpent's tail, and a goat's body; any idle or wild fancy." Certainly in their quest for what George Bush had called the "vision thing," the Freedomites had stumbled across an extravagant constitutional "Lego" kit.

Khol's conditions were not impossible for the FPÖ to meet, although technically the reforms put forward under the title "Third Republic" had become a central plank in its thinking. Formally to ditch this package would bereft the party of a "big idea." Despite this, leaders in the FPÖ had no problem with their brainchild since they always pointed out that such a revision of the federal constitution would need a two-thirds majority in the National Council (article 44 of the federal constitution). Since the FPÖ realistically could not expect such support, the Third Republic project looked stillborn. Gradually the FPÖ stopped talking about the "Third Republic" mainly because it sounded too much like the "Third Reich" and had been open to misunderstanding. The ideas for constitutional reform were therefore alternatively rebaptised "the new Second Republic" and the "free republic." Those who wanted could therefore see that positive developments in the Freedom Party were in progress.

On the question of EU membership Khol welcomed what he detected to be a more positive approach from Haider in January 1996. In a speech in parliament 15 January Haider recorded his acceptance of the decision of the Austrian people to join the European Union. Khol believed this marked a break with the F leader's previous uncompromising stance.[4] Haider, however, even before the 1995 election at a press conference before foreign journalists, had said that, as a democrat, he naturally respected the verdict of the Austrian people in the 1994 referendum on EU membership. Austria, he said then, could not be taken out of the EU without the express wish of its people. In an interview in *Kurier* 28 October 1995 he said "a decision of the people can only be changed by the people themselves." His main grumble was the Maastricht Treaty and the European superstate and that held good even in 1996. He remained a Eurosceptic as outlined in his book *The Freedom I Mean* :

> Brussels is already engaged in salami tactics to set up a dull, lifeless unitary state. But seventy years of the Soviet Union show the futility of trying to steamroller freedoms and national diversity. Political imperialism and colonialism cannot stamp out cultural identities of peoples. Any attempt to do so will boomerang and the force of nationalism will emerge. The nation state is not finished. To try to abolish it is the quickest path to nationalism. Freedom and the existence of different customs, traditions and cultures must be respected. The nation state has an important historic and cultural role and is admirably suited to the rights of citizens.[5]

The critical position of the FPÖ on social partnership had been consistently held by it and its predecessors since the war. This was nothing particularly new or Haiderised. Such criticisms had also been voiced by the Left, the Greens and the Liberals. Many others also recognised the need for wide ranging reforms in this area to make the system more modern, democratic and suitable to Europe. There was a case for a more limited role of the social partners and for more transparency. For Khol the problem was that, as he saw it, the F position was fundamentalist and sought a total abolition of the existing system. Social partnership had evolved ad hoc in the Second Republic and was not in any case mentioned anywhere in the Austrian constitution. It had been a successful hallmark of the Second Republic but was itself outside of the constitution and as the FPÖ remarked technically lacked "democratic legitimacy."

A clear rejection of National Socialism as demanded by Khol had frequently been forthcoming from Haider although other comments made by him often set the credibility of this position in doubt.

The cultural revolution called for in Haider's book was also an essential tenet of the party's thinking in weeding out Marxist thought from universities and the arts which it suspected had led to political sponsoring. The FPÖ believed that artists were at pains to toe the line to get necessary subsidies from the appropriate minister whereas those who might have contrary views were left out in the cold. According to Haider political correctness was implemented through this kind of "psycho-terror." As Haider explained in his book *The Freedom I Mean*, the ringleaders of the '68 generation were now in leading positions in society, "they wield enormous power in the fields of culture, education and the media. There they cultivate their insatiable ideological needs and conjure up artificial enemies. With their modern Inquisition they produce an intellectual climate of crippling conformism." Haider was also critical of so-called culture which depicted excess violence, sexual perversions and blasphemy but still received taxpayers' money. First steps to change this for Haider would be to dismantle the power structure of the federal theatre system, to strengthen artistic autonomy and to reduce official sponsoring of artists and performers. Party political meddling in school appointments would end: "We need well-educated children and not indoctrinated youth…. We must fight a gallant battle to defend our cultural values against the insidious forces of moral decadence."

A "revolution" was called for which did not advocate bloodshed or execution. Revolutions can be "silent" or bloodless without violence but nevertheless achieve monumental change.

THIRD PRESIDENT

Krumpendorf had soured the relations between the FPÖ and the other parties in parliament but its electoral strength, despite the losses in the 1995 election, meant it could not be ignored. This became obvious when the new parliament met in January 1996 and started the process of electing its three presidents. Recent convention dictated that these posts went to the three strongest political groups in parliament. In this case the first president should go to the SPÖ, the second to the ÖVP and the third post should then be occupied, as in the outgoing parliament, by the FPÖ. It was the third presidency which was to cause the most controversy and become a headache for the ÖVP. In the old parliament the third president had been Herbert Haupt, a member of the Freedom Party. The SPÖ supported by the Greens and the Liberals believed that the FPÖ was not worthy to provide an occupant for such a post. The three presidents take the chair on a rota basis and have the task of maintaining order in the National Council and implementing the rules of procedure.[6] They preside over the proceedings and debates, put questions to the vote and announce the results and are responsible for discipline. If necessary they can clear the visitors' gallery and call for a break in an exceptionally unruly debate.

In terms of protocol the president of parliament is in many cases the next highest in the country after the federal president. According to article 64 of the federal constitution, if the federal president is unable to fulfill his duties for more than 20 days, dies or is voted out of office in a referendum then the first president, with the two other presidents, takes over.

The ÖVP believed that the tradition whereby the third largest party should have the right to the third presidency should be upheld. It could not disguise its discomfort, however, that this was the FPÖ, a party it had placed outside the constitutional bow and whose leader frequently had to explain his position on National Socialism and the Waffen SS. The ÖVP now had to decide whether it should ostracise the FPÖ or bring it in from the cold and incur the wrath of the SPÖ, its best bet for a future coalition partner in government.

Haider tried to help the thinking process in the ÖVP by warning it that he could make life difficult for it both in parliament and in the election of Waltraud Klasnic as Styrian governor. He was supported by his anchor man, Ewald Stadler who also warned the ÖVP of the consequences of meddling with time-honoured parliamentary practice. The ÖVP kept to the recent unwritten law that the third party should provide the third presidency in parliament and accepted that the FPÖ and Haider had made a credible, acceptable statement on the National Socialist period. Khol added a rider, however, that Herbert Haupt personally could not command support from the ÖVP as a candidate for the third president. He argued that Haupt's conduct of this post in the old parliament had not been above reproach. The ÖVP had challenged Haupt's management of a vote during a special sitting in November 1995 but was not supported by the other parties. It began then to think of alternatives to Haupt for the post in the new parliament. Khol did not rule out another candidate from the Freedom Party although his socialist counterpart, Peter Kostelka, considered this would do irrevocable harm to parliamentarism and to Austria's international image. Haupt was also under fire for his participation at the Ulrichsberg meeting, the official function during which Haider addressed veterans of the Waffen SS. This was a weak criticism since many leading politicians from the SPÖ and ÖVP had gone to Ulrichsberg over the years.

Another problem with Haupt was that he had failed to make a coherent statement on Haider's analysis of the Waffen SS as a part of the Wehrmacht, remarking in a television interview that it was an "unsuccessful historical quote." In defence of his position he too referred to the former Nazis of Kreisky's cabinet and to the many ÖVP politicians with such a past. He insisted that individual and not collective guilt should be the standard to judge such questions.

A refusal to accord a party with the backing of over a million voters the third presidency ran the risk of making the FPÖ look the victim of political intrigue and persecution, a position from which it could possibly profit. It also meant that the FPÖ would be even harder to integrate and would be more troublesome and uncooperative. The acceptance of the FPÖ as having the right to the post by the ÖVP ended its isolation and gave it back some veneer of respectability which had vanished after Haider had tried to explain Krumpendorf on television but got into even deeper water.

The predicament of the ÖVP was clear since the support of the FPÖ was necessary to secure the governorship for Klasnic in Styria. The prospective governor did her bit to build bridges with the Freedom members by declaring that she was not seeking to be a history teacher. She called for a truce on the Krumpendorf dispute and an end to the incessant debate on what happened over 50 years ago. "The FPÖ," she concluded, "is a democratically elected group in Austria and as such is in parliament." Her apparently casual attitude to Austria's past horrified Socialists and Liberals in the province.

The FPÖ sought to use this bargaining position and stuck to Haupt as the only conceivable candidate for the post of third president. During the debate on the elections for the presidents, the FPÖ took up a new position in the assembly on the right instead of the centre back. This gave it seats in the front row but made it more vulnerable to charges that it was a party of the far right.

The ÖVP kept to the line that in principle the third president should go to an FPÖ member but refused to back Haupt. In the first round Haupt failed to get the necessary votes. In a second round the FPÖ put forward an alternative, the university professor Wilhelm Brauneder. He got the votes of the FPÖ and the ÖVP and was duly elected but immediately came under attack from the SPÖ for being on the far right of the FPÖ and an ideologue of the "Third Republic." Brauneder had only been in parliament just over a year and as an academic had long contributed to right-wing pamphlets.

Khol defended the support for Brauneder by revealing that he had had him checked out and had scanned his publications. Material had been provided by the Documentation Archive of the Austrian Resistance and nothing incriminating had been discovered, said Khol. "Every-thing Brauneder has published," he concluded, "is within the frame-work of the democratic spectrum."

This kind of "security check" for politicians was distasteful to some intellectuals in the People's Party, like Bernd Schilcher who commented in the *Der Standard*, 30 January 1996, that it smacked of McCarthyism. Schilcher was also irritated by an excessive concentration on the person of Haider by anti-fascist writers. "Every rubbish which comes from him is a headline. In countless colour cover stories Austrians and the world learn week after week the latest developments in the lines on his face or other parts of his body.... He has become the personification of a modern King Midas. Everything he touches turns to a political taboo." Schilcher concluded by paraphrasing former mayor of Vienna Karl Lueger who had once said, "I decide who is a Jew" with the words that the modern trend was to declare, "I decide who is an anti-fascist."

Brauneder was also supposed to have contacts with revisionist historians in Germany and with former Nazis such as Erich Schwinge, military judge under the Nazis who had given death sentences and who supposedly served on the same "academic curatorium" in Germany with Brauneder.[7] Brauneder himself vehemently denied membership of such a curatorium and repeatedly stated that he knew Schwinge by name only.[8] Brauneder further pointed out that, in all his twenty years in academic work, no one saw fit to make any criticisms that he was politically suspect or "extreme." This only came to be a topic of interest when he became a potential candidate for the third presidency for the Freedom Party.

Brauneder was controversial but probably any Freedom politician would have caused problems. Yet a concerted parliamentary isolation of the FPÖ could have brought with it exclusion from committees and have led to an even greater break-down of an already frail dialogue. The FPÖ still remained for Khol outside the constitutional bow although he accepted as "credible" the party's explanation of both Krumpendorf and Haider's TV interview on the Waffen SS. The FPÖ as a result of Brauneder's election with ÖVP support was back in the game.

Soon after, Klasnic in Styria was duly elected provincial governor for the ÖVP with the backing of the Freedom Party. The FPÖ's tactic of citing examples of liaison between the SPÖ and former Nazis, a fact reluctantly supported by the Greens, quelled the Krumpendorf fire to some extent.

In April another post of provincial governor came up this time in Salzburg. Once again the People's Party candidate, the historian Franz Schausberger, needed the support of the Freedom Party as had been the case in Styria. Whilst Khol was still persisting with his constitutional bow, Schausberger came out with a slightly more favourable analysis of the FPÖ which he described as "a democratic party which accepts our state and constitution." He added, however, that the Third Republic concept of the FPÖ was nevertheless outside the constitution. Since this was an integral part of the FPÖ armoury it was difficult to see how these two statements fitted together. Schausberger had coincidentally the previous year written a book on the Weimar period and the Nazis who had used parliament as a springboard to abolish democracy.[9] Many had seen parallels with Haider's FPÖ and detected a covert warning that talk of a new republic was dangerous and opposed to the existing parliamentary system. Just before his election, Schausberger refuted this analogy.

Khol hoped for a speedy return of the Freedom Party to the grounds of constitutionality. In order to do this satisfactorily it had to clear up the grey area in its attitude to National Socialism (something which supposedly had been done in January), as well as its attitude to social partnership and the European Union, and should come out with a rejection of the Third Republic. This was not the politics of marginalisation for Khol but "the politics of retrieval." The People's Party governor in Khol's own province of Tyrol, Wendelin Weingartner, was not impressed and believed that Haider and his clique were beyond redemption. Nevertheless Khol periodically welcomed moves by the FPÖ to come back home and saw a change of heart in its European policy. Haider had accepted the verdict of the voters on joining Europe although this was not new and did little to blunt his critique of the Maastricht EU in general.

From the beginning of the new parliament relations between the SPÖ and the ÖVP had been strained. Despite this, negotiations continued between the two on the formation of a new government. Although the SPÖ was the clear winner of the election it could not squeeze the ÖVP too hard for fear it would keel over and go into opposition. The weakness of the ÖVP as in the past proved to be its strength during the negotiations. The Brauneder election according to Khol helped the coalition come to fruition as it demonstrated to the SPÖ that the ÖVP had another option open if the Socialists were too difficult. The leader of the SPÖ parliamentary party, Peter Kostelka, disputes this interpretation since a FPÖ-ÖVP link-up would have been from the start weak and confronted with a strong SPÖ which had just emerged unexpectedly triumphant from the 1995 elections.[10]

Finally in March 1996 a revamped SPÖ/ÖVP government was formed. It aimed to tackle the problems which had existed before the election viz. the budget, the welfare state and jobs. The cabinet was smaller but no new blood was brought in – Vranitzky headed the team supported by the Vice-Chancellor and Foreign Minister Wolfgang Schüssel. In addition the SPÖ had 6 ministerial posts (Finance; Health; Interior; Social and Employment Affairs; Women's Issues; plus Science, Research, Transport, Arts and Humanities) and the ÖVP five (Agriculture and Forestry; Defence; Economics; Education, Environment, Family and Youth; Foreign Affairs). One state secretary went to the ÖVP (in the Foreign Ministry) and one to the SPÖ (in the federal chancellory). There was one non-party ministry (Justice) as in the old government. Haider described the "new" coalition as "the same as before plus cut-backs and voodoo politics."

The coalition agreed to implement a package to consolidate the budget two thirds of which would come from a reduction of public expenditure and the rest from extra revenue. Austrians were suddenly confronted with the prospect of an end to generous welfare benefits as the government tightened up on payments to students and families and sought to curb abuse. A wage-freeze for civil servants was hammered out and cuts foreseen in maternity leave and welfare assistance. The package was supposed to maintain social fairness so that so-called higher income earners came off worse. The cooperation and involvement of the social partners was also welcomed to avoid the conflicts which had flared up in other European countries. Most agreed on the need for some drastic steps not only to meet the Maastricht convergence criteria but also to end the "freebie society" which was living beyond its means. Nobody, however, really wanted to be personally affected by the measures, however necessary, in the interests of the common good. The students took to the streets in a wave of protests before parliament and succeeded in bringing traffic to a halt in Vienna. Public sympathy to begin with was surprisingly high for this action and the plight of the universities was well known. The opposition parties naturally capitalised on this discontent and denounced the package as the enemy of progress, education, women, the young and families. The FPÖ could not hope for a sympathetic hearing from students and academics but banked on the support of increasing discontent amongst the small businessmen and lower middle class who had been bracketed together by the government as the "big earners." The party denounced the austerity package for its lack of vision and for omitting to bring about real structural changes as well as for its failure in pruning administration. It accused the government parties of breaking their election promises to voters. For the FPÖ privilege, waste and corruption went untamed in institutions such as the Federal Reserve Bank where the so-called "fat cats" were reported to enjoy subsidised meals and cheap accommodation.

The SPÖ/ÖVP not only agreed to form a new coalition but succeeded in drafting federal budgets for two years (1996-7). This necessitated a mammoth session in parliament to push the savings package onto the statute books. All the opposition parties complained at the lack of time to scrutinise the proposals and talked of an abuse of parliamentary government. One Liberal member of parliament noted that the documents concerned weighed a substantial 22.5 kilos impossible to intellectually digest in the short time available. The coalition was determined to get its proposals through intact. Renegotiation or major concessions for one group would have led to a flood of demands from others affected. The Liberals withdrew from committee deliberations on the bills, almost 100 in all, as a protest against the government's inflexibility. The government with its new two-thirds majority could steamroller through changes in constitutional legislation in record tempo and disciplined ritualistic voting ensured a safe passage for the new coalition's proposals. It was doubtful if most members of parliament really knew what they had voted either for or against but by April 1996, it was "mission accomplished" for the new coalition.

All Change

After the 1995 election and the Krumpendorf affair, the fascination of Haider for the public and journalists receded. The two main parties started negotiations on forming a government and began a complicated programme for long-term budgetary reform. Details on hospitals and social welfare policy did not lend themselves to the sound-bite attack of the populist politician. Haider left the hard graft of this work to the would-be government and kept an uncharacteristically low profile. His comments on the Waffen SS had not enhanced his standing or furthered his declared cause to reform Austria. Many asked exactly what kind of future renewal Haider had in mind. In short, Krumpendorf looked like an own goal which catapulted the party back to square one. This called for a time of reflection and a reassessment of long-term aims and strategy.

At a New Year conference in Linz in 1995, the FPÖ had moved away from the traditional party concept to a kind of electoral movement for the presumed election to be held in 1998. Already by the summer Haider had second thoughts about the wisdom of the "F movement" and the "FPÖ," as the more conventional identity tab, was rehabilitated. The idea of holding primaries on the US model before an election was not advanced enough by the sudden election of December and the party adopted the standard approach it had used for the last decade. It was doubtful anyway if the primaries and electoral conventions were really suited to the Austrian party. Many of the party's voters were happy to put a cross in the election booth but were shy about being associated with it in public under the full glare of the media.

The election in December was a financial disaster as well as political setback and the FPÖ was forced to reconsider its priorities. This led to some key personnel changes in March 1996 and Gernot Rumpold, who bore the blame for bungling the late claim for election expenses, was given a new job concerned with advertising. He was replaced as party manager by Karl Schweitzer, environmental spokesman and then member of the European Parliament. At Linz in 1995 the party had done away with the posts of general secretaries, a move seen later as a mistake. Schweitzer had represented Burgenland in parliament since 1990. The appointment of Schweitzer was interpreted *de facto* as a step towards the reinstallment of a kind of general secretary. Schweitzer saw one of his main tasks as being to improve the relations between the central party and the individual provinces. He was also to take over the job of running the Alliance Office (*Bündnisbüro*) which aimed to promote links with citizens. Many of the ideas floated at Linz had never become airborne and a year later were put back in the hanger to wait for a suitable following wind. These kind of aerobatics incurred criticism from some in the FPÖ (or F) who failed to detect a clear position on the idea of a party (and/or movement). The voters of the FPÖ, however, had only very loose links with their party. The idea of the Alliance movement had been to offer these people another option besides full membership so that they could receive information material and invitations to events. The project was not rated a special success in the first year of operation and it seemed people preferred either to opt for full membership or nothing at all beside the commitment on election day. For some it had been an interesting gimmick and had kept the organisation-movement in the headlines. For others it had simply been a confusing experiment.

The 1998 project was shelved but not buried. The problems which had led to an early election had not disappeared and it was premature for the FPÖ to totally abandon its declared course. The FPÖ, it seemed, could shift around on policy and organisation much easier than its main rivals. The SPÖ had evolved as a mass membership party for over a century and could not easily dispense with it despite its lack of suitability in the electronic age. The FPÖ had no such rigid structure and its supporters were a heterogeneous group. It preferred therefore to retain a flexible response on organisation, policy and personnel. The leader of the delegation to the European Parliament, Susanne Riess-Passer, took over as head of staff in Haider's office. Press spokesman Peter Westenthaler was put in charge of a new communication office. The frequent switches in personnel were explained as a deliberate policy of "job rotation."

The centre of gravity of the party's work shifted more to parliament which was more economical to manage since it could rely on some public funding, and gave Haider the chance to steal the headlines as floor leader. He was supported in this task by Ewald Stadler from the westernmost province of Vorarlberg who, because of his sharp and biting attacks, had earned the nickname of the "Doberman dog" from his opponents. Stadler was born in 1961 of a staunch conservative, Catholic family. Some of his relatives had suffered under National Socialism and had even been killed while fighting in the resistance in the last weeks of the war. He entered parliament in 1994 and soon sat as number two along with Haider on the party's front bench. He had stood by Haider through the difficult days following Krumpendorf and had built up a reputation as a hardliner. Stadler had come up through the student duelling societies and was proud of this "character-building education." He himself had fought four duels and had the scars to prove it. In addition, the parliamentary group appointed a corps of ten policy spokesmen to spearhead the attack on the government.

The election had taken the wind out of the sails of the party, but it was clear that many of the old problems such as privileges for politicians, reform of the economy, jobs and freedom from party cartels were still there. The new government was working on a budget which would hit students, pensioners, and women without bringing in the ambitious structural reforms promised before the election. In many ways this should have been the hour of the opposition but after the 1995 election these parties looked disoriented. Since many of the problems were the same, the solutions remained unaltered. The Freedom Party found itself repeating its position on reducing non-wage labour costs, calling for less taxation and a lean state, but ran the danger of sounding stale. When it developed new ideas these often seemed for show and not carefully thought out. The aftermath of Krumpendorf left the party with some hard thinking to do.

FORTY YEARS OF THE "FPÖ"

Easter 1996 saw the fortieth birthday of the Freedom Party of Austria. Commentators were unsure whether it was still the FPÖ or simply F, and whether it was once again a party or the new "movement." For Haider it was all these and much more – a force to change Austria and to move other parties and politicians in this direction. A celebration documenting the early history and evolution of the party proved an embarrassment. The years pre-Haider were not considered the most spectacular in the party's history and the last years had seen the departure of some leading figures, such as Heide Schmidt & Co., who had become *persona non grata*. Despite this an anodyne video was put together concentrating predictably enough on Haider's successes since taking over in 1986. Around 600 guests turned up to the anniversary festival in Vienna, although previous leaders of the party were conspicuous by their absence. Friedrich

Peter seemed accountably peeved he had not been invited even though he was no longer a party member and had boycotted the last invitation to such an event. Dissidents such as Schmidt and Frischenschlager were shrugged off by Haider for their "betrayal."

Haider himself appeared in a new designer outfit for the occasion and closely cropped hair, apparently intended to emphasize the sporty side of the movement which had to be fit for the coming electoral challenges. According to political analysts the new image was targeted to appeal to a lower middle-class sense of sobriety. To pull in more votes the party had to go beyond its "prole" image cultivated to woo the working class. In a television interview (*Report*, May 1996) Haider defined the party as "centre right" which stood for some conservative values. He described the FPÖ as a classic middle-class party which wanted more market economy, less tax and less state. Elsewhere Haider would talk of the FPÖ as a kind of people's party, an alliance for reform representing those who wanted innovation. He appealed to the little man who was losing out and sought security against large international concerns whilst supporting the modernisers who wanted to get on free from an over-regulated and protected system.

In the booklet (*40 Years of the Freedom Party of Austria*) accompanying the show, Haider recognised the difficulties of the path ahead:

> Our way to a free and open society is not easy and will not be simple in the future. Those who demand a renunciation of power will be fought by those who consider this power as their own property. Whoever like us defends the basic freedoms of citizens will be the enemy of those who abuse these freedoms. Whoever like us stands for direct democracy will meet with the resistance of those who regard themselves as the guardians of the state and its citizens. Whoever like us has as their aim an open society will be seen as a threat by those who for decades have set up a system of political dependence.

The federal president, Thomas Klestil, was the only Establishment notable to show up and even he left early because of a pressing schedule. In his speech the president stressed the importance of the FPÖ as an opposition party for democracy and came out against all strategies to marginalise the party: "Those act undemocratically who deny others integration." Klestil recognised the role of the party as a reform movement with courage to seek changes where power structures had become too inflexible. Many of Klestil's own speeches had criticised the dusty nature of politics in Austria. In his television address to the country in October 1994 Klestil remarked that,

> Some political structures have not kept pace with necessary developments in democracy or have got stuck in political rituals. Many Austrians no longer have any understanding for the existing spheres of influence and power, for bureaucratic and institutional immobilism, and non-transparent decision making procedures. They no longer find the omnipresence of parties in tune with the times and do not understand why the occupation of important functions in our democracy – from school director to supreme court judge – is still dominated by party politics. They want more objective cooperation and less *Proporz*.

Klestil added a word of warning before those assembled to celebrate 40 years of the FPÖ, "No one has a monopoly of truth and decency or of patriotism and democracy." A state without parties was inconceivable but "the state should never become the possession of parties and party interests." Chancellor Vranitzky criticised this speech for failing to address the contacts of the FPÖ with rightist groups.

The electoral success of the party under Haider was indisputable. Despite this, power remained elusive and observers noted the isolation of the party both within Austria and on the European stage. The European Parliament condemned (April 1996) Haider in the following resolution: "...that racist parties constitute the crystallisation point for xenophobia, racism and anti-Semitism in society and it is necessary to ostracise them and isolate their political leaders, such as Mr. Le Pen and Mr. Haider, in the Union in order to combat racism and anti-Semitism." Only a third of the 626 members of the parliament were present. Socialists, Communists, Greens and Liberals voted for the motion (127) against the conservative parties including delegates from the ÖVP (88). In the European Parliament, the Freedom Party was unable to find supporters to join a political grouping. The party was successful but in the wilderness with little prospect of acceptance. Haider remained undeterred and in his message to the birthday guests concluded: "the political game is in the future unthinkable without us." Haider's strategy was to work on the other parties so that they would react to his agenda. He was satisfied that on many issues such as immigration, privileges for politicians, the economy and defence the established parties had adopted his tone and responded to his tempo. This was no substitute for real power but proved that "isolation" could be, if not splendid, at least productive.

One problem for Haider in the 1995 election was that the traditional issues which had favoured the FPÖ, such as privileges for politicians, security and immigration, had been of secondary importance. This did not mean they had disappeared and in due course they resurfaced to dominate the political agenda to the benefit of the Freedom Party.

THE TIDE TURNS

At the beginning of June 1996, provincial elections were held in Burgenland in which the FPÖ increased its share of the vote from 9.8 percent to 14.6 percent. This was put down to disillusionment with EU membership and as a protest against the government's savings measures.

In July former left Socialist Josef Cap opened up a controversial debate by seriously proposing an examination of NATO membership, an abandonment of neutrality and the introduction of a professional army. Some queried ironically whether he had gone over to the Freedom Party since such ideas had long been official policy of this party. On the question of immigration some Socialists too advocated a hard-line approach especially as many firms were struggling and job losses were constantly in the headlines.

In the summer of 1996 the savings package passed by the government was beginning to show up on wage slips. It now became palpably clear that most people would have less take-home pay and would have to rethink expenditure priorities. Prescription charges went up and other increases were in the pipeline to rescue the parlous state of the health service.

Coincidentally a controversy exploded on politicians' pay and financial perks which put the government and especially the ÖVP in an embarrassing position. The ÖVP member of parliament and head of the party's league for workers and employees, Josef Höchtl, was publicly outed by university authorities where he was supposed to teach. In addition to his pay as member of parliament and league boss, Höchtl had pocketed for over twenty years the equivalent of twice the average wage for lecturing but had never shown up at the university for classes. Legally this was in order according to the regulations governing civil servants but the public had little sympathy for Höchtl, although he failed to see why he alone should be singled out when hundreds of others were also getting paid for nothing. This story was particularly irksome for many students and junior lecturers who were suffering from government cut-backs. More culprits were to be found in the other parties but the main damage was inflicted on the coalition. Night after night politicians from all parties were obliged to go live before the cameras and disclose their earnings from different sources and justify what they did, if anything, to earn the money. In an embarrassing moment the Socialist president of the National Council rushed back to the studio to correct his statement since he had inadvertently overlooked 40,000 Schillings per month. Höchtl finally renounced claim to his money and resigned as leader of the ÖVP workers' league.

The entire spectacle put politicians in a bad light when their standing was already at a nadir. The coalition decided to act and drew up a scheme which would get rid of the anomaly so that politicians would not be able to get extra pay from public funds for doing nothing. At the same time, however, expenses were increased so that in most cases those concerned were financially better off than before the reform. A reform of the reform was therefore hastily worked out to be rushed through before the parliamentary recess. This needed a two-thirds majority in parliament which it just scraped since it met with scepticism from deputies in all parties. It was generally acknowledged that the entire package was a sticky tape job until something more substantial could be worked out. It dealt with the tip of the iceberg (the Höchtl case) but left the rest untouched. Yet another problem had been shelved and substituted with a second-best pushed through at the last moment in reaction to public odour. This showed how dangerously out of touch politicians had become with the needs of ordinary people.[11]

The Freedom Party also had deputies with more than one paid job, financed with public money.[12] It claimed, however, that their people actually did some work and that anyway since January 1995 those earning over 60,000 Schillings per month net paid the excess into a social fund. Only 16 politicians came into this category and with few exceptions they paid the contributions. The other parties suspected that this was some illicit party financing and called for details of the recipients of the fund. This the FPÖ did, but omitted personal details of individuals for reasons of data protection. According to Gilbert Trattner, FPÖ member of parliament who administered the fund, payments had been made to individuals and institutions in need such as a ten year old girl who had suddenly become an orphan. Money had also gone to a spastics' association, to someone suffering from leukemia, and to a hospital for equipment. A further donation had been made for blind children, for wheelchairs, for a children's playground, and for those in special need as part of a Christmas '95 project. The fund itself looked clean but it was suspected there were ways to dodge paying into it.

THE SPECTRE OF TERRORISM

The new star in the party hierarchy, Karl Schweitzer, soon came under attack for his associations with right wing extremists. Every time, it seemed, when the party was making headway it would be confronted with this uncomfortable accusation. In 1992 in Burgenland a Jewish cemetery had been desecrated and daubed with swastikas and "Sieg Haider" slogans. In August 1996 the suspected culprits were apprehended and described as members of the extreme right fringe. Not only this, but it was also revealed that they had been schooled by Schweitzer, a teacher by profession. They had been encouraged to become politically active in the party by Schweitzer and one of them had even stood as a candidate on the local party list. Schweitzer originally had some problems in remembering his associations with the men concerned despite the fact that they all came from a small community. Under pressure in a television interview he finally stammered that it was not possible to look inside all those who approached the party with an interest. Austrian radio and television later produced a tape of a meeting in which Schweitzer had referred to the men concerned and from this it was obvious that he had been more closely involved with them than he had at first acknowledged.

The news was a gift to the other parties seeking to corner the FPÖ into a difficult position before important elections in Vienna and for the European parliament. Calls were made for Schweitzer to resign and he seemed to be made guilty by proxy for this act of vandalism. Haider defended his party manager, with some twisted logic, by refusing to dismiss him so long as the Interior Minister, Caspar Einem, remained in office. The episode came at an awkward time for the FPÖ which was busily polishing up its image for the autumn elections. Schweitzer survived until after the elections but was replaced in November 1996 by his predecessor Gernot Rumpold in another example of F job rotation policy. (Schweitzer was given the new task of EU coordinator.)

Many authors had repeatedly commented on the links of some FPÖ politicians with the extreme right fringe in Austria.[13] The FPÖ was not helped by a periodic wave of letter bombs and other incidents of violence. These often occurred at a time when the party was going for the votes of the respectable middle-class and such acts of terrorism could only damage its fortunes. The letter bombs claimed some prominent victims, including the then mayor of Vienna Helmut Zilk (December 1993) who suffered permanent injuries and a mutilated hand. The authorities seemed powerless to track down the culprits and the waves of bombings continued at various intervals. The targets for these attacks were mostly those who worked to integrate foreigners in Austria or who were known for promoting this cause. The perpetrator(s) issued written documents calling itself the "Bajuvarian Liberation Army" which were assumed to be penned by xenophobic right extremists. Little tangible emerged from the police inquiries and years elapsed between the first wave and later bombs which became increasingly sophisticated. The general director for public security, Michael Sika, admitted in an interview with *Die Presse* in January 1996 that the original assumption that the culprits came from the right was too simplified and made under pressure from the media and some politicians. He added that to have declared in the beginning that the bombs were not the work of neo-Nazis would have meant that "I would have been publicly hung."

In February 1995 four members of the Romany community were found dead after a bomb exploded in Oberwart, Burgenland. The device was attached to a sign which read "Romanies back to India." Shortly after in Stinatz (an area in Burgenland with a Croatian population), an explosive severely injured another victim. The automatic assumption was that the culprits were most likely to be found on the extreme radical right. But for the Freedom Party this was too simplistic and it was highly critical of the police investigations for starting from this premise.

In addition to this suspected right wing terror there were also isolated incidents of leftist extreme activity. In April 1995 an attempt to blow up an electricity pylon in Ebergassing which would have left Vienna without electricity failed and the culprits themselves lost their lives in the process. The bodies of two men known were found nearby but a suspected "third man" could not be traced. The men were linked with the far left "anarcho" scene in Vienna which produced journals such as TAT*blatt* which was even receiving subsidies from the state. The FPÖ found this leftist activity a mixed blessing since it took the heat off their own problems and links with the right. The party unearthed an extensive far leftist network with contacts with the Greens which sought to undermine the republic. It included so-called experts on right wing extremism who even were subsidised to give talks to school classes in educational programmes. Increasingly the main target for the FPÖ was the new Socialist Minister of Interior Caspar Einem who, it was claimed, had ties and sympathies with the far left and personal contacts with the men behind Ebergassing. Einem had an unconventional background – born into the nobility and a descendant of Bismarck, he had once worked as a probation officer and had even made personal donations to the TAT*blatt*. He claimed he wanted to help drop outs, drug addicts and those in trouble to re-enter the community and become law-abiding citizens. He was criticised variously for naiveté, idealism and incompetence but became the hero of the left in the SPÖ. He was elevated to a kind of anti-Haider figure and supported by the Greens. The ÖVP was unhappy about Einem but refused to back the FPÖ in passing a motion of "no confidence" in the minister in parliament. Einem was personally affected by these terrorist attacks when a letter bomb was posted to his step-mother in Vienna. The letter was sent to the wrong address and the bomb squad alerted and no one was injured.

Before the European elections there was another bomb threat scare and long coded letter was deciphered by the authorities from the "Bajuvarian Liberation Army." It was full of the usual anti-Slav hatred and mentioned leading journalists and the chancellor Vranitzky as future victims for its sinister operations. It also acknowledged responsibility for the terrorist acts of Oberwart and Stinatz.

The election campaign started under this spectre of terrorism but passed off without any act of violence. It created some nervousness but did not damage the electoral position of the Freedom Party.

FIGURE 6

CONSTITUTIONAL BOW

SECOND REPUBLIC
(1920 Constitution as revised in 1929)

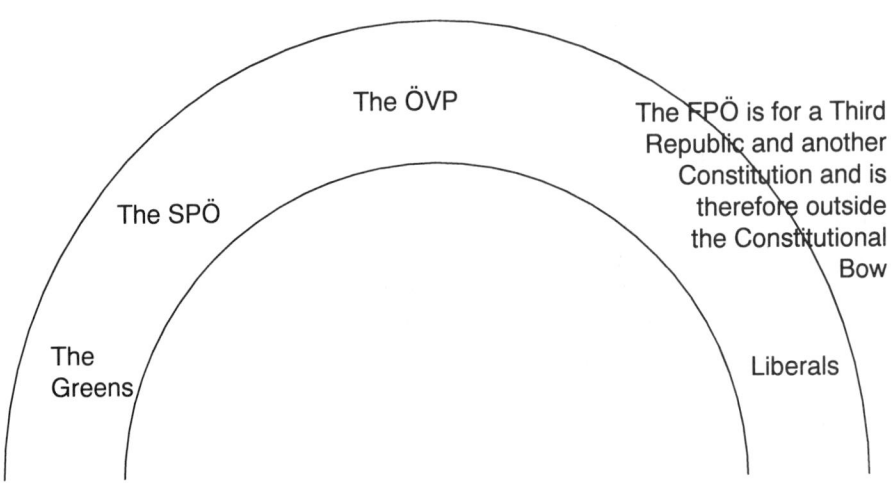

Source: Khol, A. *Pressestunde*, May, 1995.

NOTES

1. See A. Khol, "Die FPÖ im Spannungsfeld von Ausgrenzung, Selbstausgrenzung, Verfassungsbogen und Regierungsfähigkeit," in Khol-Ofner-Stirnemann, eds., *Österreichisches Jahrbuch für Politik '95* (Munich: Oldenbourg, 1996), pp. 193-221.
2. Interview with *News*, 44/95.
3. Interview with Pelinka, June 1996.
4. Interview with Khol, May 1996.

In a television interview in November 1996, Khol still argued that the FPÖ was outside the constitutional bow because it rejected the European Union, still had an ambivalent position on National Socialism and sought a new "Third Republic." It was, in short, unfit to govern. See also *Der Standard*, 25 November 1996, in which Khol once more complained that the FPÖ was against social partnership and the Austrian nation. He criticised the party for subsidising a festival of the right wing student duelling fraternities and charged Haider for his rabid rhetoric against the Austrian republic, often made abroad. By 7 January 1997 (*Profil* interview), Khol saw some positive signs in the FPÖ and Haider, since he noted that the phrase "Third Republic" had been dropped, a clear break had been made with "neo-National Socialism," and there was no talk of leaving the European Union. Khol even detected a positive change of heart from Ewald Stadler, Haider's right-hand man. This, despite the fact that Stadler had attended the pan-German students' meeting and had shown no inclination to revise his basic opinions. Khol's sudden re-assessment was made in the midst of a crisis in the coalition which obliged the ÖVP to seek temporarily the support of the FPÖ in parliament.

5. *The Freedom I Mean* (New York: Swan, 1995), pp. 82-83.
6. For a commentary on the election and duties of the presidents in the National Council, see K. Atzwanger and W. Zögernitz, *Nationalrat-Geschäftsordnung* (Vienna: Manz, 1990), pp. 44ff. and pp. 66ff.
7. *Kurier*, 16 January 1996.
8. Interview with President Brauneder, July 1996.
9. F. Schausberger, *Im Parlament um es zu zerstören – die Nationalsozialisten in den österreichischen Landtagen 1932/33* (Vienna: Böhlau, 1995).
10. Interview with Kostelka, July 1996.
11. The coalition tried to grapple with the problem and worked on an "incomes pyramid" together with the Greens and the Liberals in parliament. The FPÖ refused to countenance any model which paid politicians more than 60,000 Schillings net a month and was not involved in the negotiations. The four party agreement proposed massive cuts for the federal president and some other politicians but also financially upgraded the posts of federal chancellor, vice-chancellor, first president of the National Council and the parliamentary floor leaders. The FPÖ was disgusted that such increases were contemplated at all in a time of cut-backs and hardship. Like the immigration issue, trying to tackle the question of how much politicians should be paid was to be a "no win" situation for the government. It could never compete with Haider and any reform measures it passed were hardly likely to satisfy an increasingly critical public.
12. One prominent case was the third president of the National Council, Wilhelm Brauneder who was paid for this job plus a university professor's post and also received an income for being a municipal politician. The last job he gave up under pressure from the opposition and the media, but also from the leadership in the FPÖ. In July 1996 Haider declared that he earned 72,895 Schillings per month and paid the amount above 60,000 into the party's social fund.

The media were never tired of asking how Haider managed to have such a lavish lifestyle on this income. In the summer of 1996, for example, Haider paid around 200,000 Schillings for a course at Harvard University. For Haider, however, this was no luxury but simply his summer "holiday."

13. R. Gärtner, *Die ordentlichen Rechten – Die "Aula," Die Freiheitlichen und der Rechtsextremismus* (Vienna: Picus Verlag, 1996).

12

EURO ELECTIONS '96 AND AFTERMATH

Since Austria joined the European Union (EU) at the beginning of 1995, Euroscepticism had grown in strength. The enthusiasm for membership shown in the referendum result of June 1994 had vanished and given way to sober reality. The parties which stood to gain most from this frustration were the Greens and the Freedom Party. Elections to the European Parliament for the Austrian delegates were necessary and were finally arranged for 13 October 1996.

The problem for the FPÖ was finding a dazzling candidate to lead the list. The party had sent in the interim period five deputies to Strasbourg headed by Susanne Riess-Passer who in May 1996 returned to Vienna as chief of staff to Haider. The other parties also had problems finding candidates who could pull in the votes. The SPÖ plumped for a municipal politician with European aspirations, Hannes Swoboda, and the Liberals wheeled out Friedhelm Frischenschlager, former FPÖ Defence Minister. The ÖVP showed most inspiration by calling on Karl Habsburg, grandson of the last Kaiser, and Ursula Stenzel a former television news presenter. This move was initially resented by the party rank and file who felt by-passed by Schüssel and his strong man Khol. The FPÖ preferred to concentrate its fire on the weaknesses of the government and made the most of the growing unpopularity of Europe. It could afford to delay a decision on candidates and made a virtue, as so often, of necessity.

Anti-Europe feeling potentially provided the FPÖ with an array of ammunition to attack the government. The line of the party was not especially new or breathtaking but followed the position taken before the referendum. According to this the Freedom Party wanted a confederation of sovereign states. The arguments were much the same as the Referendum Party in England, funded and founded by Sir James Goldsmith: anti-Maastricht, anti-son of Maastricht and anti-Euro (a common currency) unless sanctioned by the people in a referendum. The FPÖ reckoned on the support of pensioners and small savers, who feared they could lose out and have to cope with backdoor inflation through "rounding up" of essential food and fuel prices. The party also sought to make the European Union more democratic, open and closer to the needs of citizens.[1]

The FPÖ opposed the idea of the "occupied fields," a doctrine which regarded centralisation as irreversible once in position. It wanted to see where possible the return (repatriation) of decision-making powers to the individual countries and the building of loose coalitions for specific purposes on integration where necessary. As in domestic politics, the FPÖ stood out against corruption and fraud and demanded tighter controls on the flow of subsidies for dubious causes. It exposed Austria's teething troubles as a newcomer in the EU and the difficulties it had in speaking with one voice in Brussels. It criticised too a personnel policy which too often smacked of the same notorious party patronage which operated in Vienna.

Part of Haider's critique of the European Union was also reminiscent of the leftist charge that the EU was a rich man's capitalist club. He bemoaned the influence of the multi-nationals and took the side of the ordinary people against these insidious power structures. He also linked the highest unemployment in the Second Republic with Austria's membership in the EU.

"VIENNA MUST NOT BECOME CHICAGO!"

Haider finally announced his list of candidates to fight the European elections and surprised friend and foe alike by putting forward the author Peter Sichrovsky, former critic of the FPÖ leader and son of Jewish parents who had fled the Nazis before the war and went to England. His writings had dealt with National Socialism and its consequences. He did not believe however that all survivors of the Holocaust were saints nor that all SS members were murderers. His decision to stand on the Freedom Party ticket met with hostility and outrage from representatives of the Jewish community in Austria and abroad. He was branded as a traitor and seen as Haider's "Jew boy" brought in to make the "F movement" respectable. Friends who had known and worked with Sichrovsky confessed to being baffled by this step and warned against his being used by Haider who, they said, was keen to have a Jew in his movement at all costs. Sichrovsky was accused of needing the money, of being desperate for publicity and at best was seen as a crass opportunist. Sichrovsky himself was unperturbed by this reaction and was even amused that most newspapers avoided describing him as Jewish and euphemistically referred to his "Jewish origins."

Sichrovsky dominated the headlines although the party list was headed by the young University Sports Scientist and former trainer of the Austrian ski team, Franz Linser from Tyrol, already representing the party in Strasbourg. Nominations also included a journalist Hannes Kronberger who had once worked for a left wing magazine and was preoccupied with "green" environmentalist issues. Many in the FPÖ suspected his heart still lay with the left but tolerated the entire list for the sake of a good publicity stunt.

Sichrovsky, too, was reckoned to be a former left liberal and had worked as a journalist for the Viennese daily *Der Standard* and the German magazine *Spiegel*. He preferred to describe himself as a liberal conservative and refused to countenance membership of the FPÖ. He lived in Chicago with his American wife, in the city which paradoxically the FPÖ had brought into its campaigning. Elections were scheduled for the Vienna municipality on the same day as the European election and the FPÖ slogan in the capital was, as once before, "Vienna must not become Chicago!" Former US president Jimmy Carter on a visit to Vienna, after seeing the placards, asked "what's that about Chicago?" The mayor of Chicago, Richard M. Daley, asked for an apology from the Freedom Party, adding it was as if Americans would say Vienna is full of Nazis. The Vienna FPÖ refused to apologise to Daley for its campaign and recalled the Waldheim affair when those abroad had tried to influence an Austrian election but to no avail. It insisted on the relevance of the slogan pointing out that the murder rate in Chicago is seven times higher than in Vienna. The American Ambassador in Vienna, concerned that Chicago was equated in the minds of Austrians with Al Capone gangsterism, asked Haider to personally intervene to remove the offending posters which despite a sympathetic reply remained in position. Further protests came from the US embassy and, just before election day, new posters went up in the city.

Sichrovsky declared he would continue to live in Chicago and commute to Strasbourg for plenary sessions in the parliament at his own expense. He was apparently attracted by Haider's mission to break Austria's old structures. His candidature was bizarre in many ways. Sichrovsky had once personally attacked Haider especially after his comments on Hitler's employment policies. Both he and Haider now brushed this casually aside. As Sichrovsky saw it,

> I never criticised the party. I criticised Haider himself for using the words he did on the Nazis' employment policies and for doing this in a political debate because he broke with a taboo in our society never to make any positive comparisons with the Nazis. I never wrote anything against the Freedom Party. I always respected Haider and it was almost that I was angry because he destroyed and wasted the chance to do something different compared with the others. I thought at the time he had destroyed his political career through a stupid mistake. But he came back and he's much bigger than before.[2]

For Sichrovsky Austria was lagging behind other democracies such as the USA especially in the fields of media and culture. He believed Haider had genuinely changed and was fond of relating how the F leader had taken his daughters to the Holocaust museum in Washington.

The FPÖ could claim it was open for former leftists and for intellectuals with whom it had a vexed relationship. Most academics however were not impressed and the political scientist Anton Pelinka wrote in *Der Standard* that "the friendship between Jörg Haider and Peter Sichrovsky has the intellectual charm and the moral quality of the Hitler-Stalin pact." Haider's list had the desired effect of capturing the headlines. The discussion was dominated by the appearance of someone "of Jewish origins" on the FPÖ ticket. Curiously few commentators chose to reflect on the suitability or otherwise of Sichrovsky as a candidate for Europe. The rank and file in the FPÖ mumbled misgivings about the unconventional choice of candidates for the European parliament which was loaded in favour of the newcomer with little experience in the party. In the ÖVP too the ordinary members felt cheated by Schüssel's list which gave prominence to the grandson of the last Habsburg emperor and a non-party member journalist. Parties increasingly tended to plump for side-door entrants with little practical experience in politics but who nevertheless were well known.

BLACK AND WHITE

After a period in the doldrums Jörg Haider and his Freedom Party were beginning to show their old ebullience. Members were especially amused by the difficulties in the rival Liberal Forum. For the Vienna elections the Liberals had nominated an opinion pollster,

Wolfgang Bachmayer, who decided to try his hand at politics. He was soon to find that the business was even more unpredictable than he had been used to in his old job. Even when he announced his candidature he was accused of having played tennis with Jörg Haider, damaging enough for a budding liberal but just before the Vienna election he was confronted with some embarrassing statements he made in 1992 at a Freedom Party symposium. According to the records of the conference, Bachmayer had endorsed some suspect anthropological theses with the remark that he had once flicked through a scientific work and learnt that white babies scream at the sight of people's faces with another, i.e., dark, skin colour but reacted normally to white people. He concluded that there was obviously here some anthropological laws of nature at work. Confronted with this on television, Bachmayer firstly denied the accuracy of the quote, then stammered he could not really remember and then pleaded schizophrenia arguing he had presented a position not as a politician but as an academic. For Heide Schmidt and the Liberals whose *raison d'être* was racial tolerance and integration the revelation was a stunning blow. Bachmayer resigned as the leading candidate of the Liberals confessing that he had then lacked the political sensitivity which he now possessed. Schmidt tried to explain the incident away by claiming that her party was capable of learning from mistakes in contrast to the others. She was clearly distressed by the fact that her front man had found nothing wrong with what she considered blatant prejudice. The case introduced an unusual slant into Austrian politics since colour of skin, unlike in America, was not normally an issue. Anti-Semitism and xenophobia were more traditional sores in Austria.

Schmidt had scarcely left the press conference to announce Bachmayer's resignation when she was confronted with another major setback linked again with the Freedom Party she so detested. One of her own parliamentary deputies, the little known Reinhard Firlinger, suddenly told Schmidt and her colleagues that he could no longer work with them and would be crossing the floor that day to join the FPÖ parliamentary group. The statement came like a thunderbolt to an already depressed Schmidt. Firlinger confessed to having been for some time "fascinated" by Haider. He also saw Sichrovsky's candidature as a positive sign that the FPÖ was open for liberals and had taken advice from Sichrovsky while making his decision. Sichrovsky told him: "Either you lie to your children and have a nice life or you're honest with them. It depends if you want to carry on in a party you don't agree with." Firlinger apparently had for some time felt unhappy that the Liberals were concentrating too much on peripheral issues, such as fighting the crucifix in schools and promoting homosexuals, instead of economic policy. He was not unique in this but could claim some fame in being the first deputy to leave a parliamentary group to join another. Previous dissidents had stayed in the chamber as so-called "wild" deputies unattached to any group. When Schmidt had left the FPÖ in parliament she did not join another group but founded her own Liberal Forum. With this action Firlinger boosted the number of FPÖ deputies to 41 which entitled it to claim almost 5 million extra Schillings per year in the current legislative period.

Firlinger found Haider "fascinating" since he articulated original topics such as the need for constitutional reform to deal with the problem of the party state. The new recruit had his political roots in Lower Austria and had always tended to see himself as a "bourgeois" liberal with conservative leanings. He was clearly disappointed with the way Schmidt had hijacked the party for populist pseudo-leftist policies. Firlinger was particularly scathing of the Liberals for their impotence in opposition. An example of this was the acquiescence of the Liberals in a reform of the parliamentary rules of procedure. The Liberals went along with this, according to Firlinger, even though they were hit themselves, to spite the Freedom Party who also lost out. He was irritated by this blind hatred of another parliamentary group which blunted effective opposition. Firlinger complained that the main priorities in the Liberal parliamentary group were dictated by Schmidt who had little interest in listening to critics.[3]

MAASTRICHT TO WATERLOO

The elections to the European Parliament (see Table 10) and for the Vienna municipality in October 1996 proved to be an unmitigated success for the Freedom Party and a disaster for the Socialists. A jubilant Haider announced "Waterloo" for the Maastricht addicts in the Establishment and restated his bid for chancellorship by the end of the century.

The People's Party too had good cause for celebration and emerged as the strongest party in the European elections. Schüssel's dream to become "Number 1" had come true at least for the

Strasbourg parliament. This was a great psychological boost to the morale of a party which had trailed the SPÖ since 1970. This was the only sweetener for the ÖVP however which suffered total defeat in the elections for the Vienna town council. It also lost a seat in the National Council to the Freedom Party after a by-election held on the same day. The People's Party had benefited overall from the so-called "Stenzel effect" of its leading candidate, Ursula Stenzel, a non-party member. After a shaky start she developed a personal and professional campaign. Analysts reckoned that without Stenzel, the FPÖ could have beaten the ÖVP into first position.

One of the most important issues during the campaign was employment. Many workers felt the SPÖ was failing on this front and no longer fought for jobs coming increasingly under threat. The less well-off and working-class families were also smarting from the impact of the government's austerity package. Although the SPÖ had negotiated this in tandem with the ÖVP, it was Vranitzky's party which took the blame for unpopular measures (see Table 11 and Figures 7-11).

The European election consolidated the trend to three middle-sized parties in place of the old red/black hegemony. The municipal elections in the capital ended the legendary era of "red Vienna" and saw the collapse of the SPÖ's absolute majority (see Figures 12 and 13).

These results hit the Socialists particularly badly and calls were made for Vranitzky to step down either as party leader, chancellor or both. The Freedom Party had, according to political analysts, established itself as a new type of workers' party. The SPÖ which had raised anti-Haiderism to a kind of party programme saw its own people defecting in vast numbers to the arch-enemy. According to psephologist, Professor Fritz Plasser, 50 percent of blue-collar workers turned out for the FPÖ compared with only a quarter for the SPÖ in the European election. This was a dramatic change around compared with the days before Haider's leadership. In 1983 the FPÖ managed only 3 percent of the workers vote; this jumped to 34 percent in the 1995 election. Plasser commented on another major breakthrough for the Freedom Party, traditionally favoured by male voters. In the European election 33 percent of working women voted for the FPÖ, a higher proportion than for any other party (see Table 11).

On election night it was Vranitzky who looked dazed and confused, the same Vranitzky who only 10 months before had basked in the success of a general election. However as some remarked at the time the votes of December 1995 were "on loan" to the SPÖ and given in the belief that the party would stand by the "little man" and safeguard the welfare state. In the follow-up to the European elections the SPÖ had failed to respond to the plight of the workers threatened by an uncertain future and job losses. It seemed too arrogant to listen and ineffectual in coalition. Throughout Austria the story was the same. The Socialist vote in its strongholds, in towns and industrial areas crumbled. The Freedom Party advance mercilessly continued. It made inroads in Vienna's municipal apartment blocks, the pride of "red Vienna" where once a large vote for the FPÖ would have been unthinkable. As the SPÖ was quick to point out, by way of consoling itself, this had not been a general election and many had not bothered to vote. Even so the extent of the Freedom Party gains at the expense of the SPÖ could not disguise the fact that the sand was running out in the hour glass. Several crisis meetings followed in the SPÖ headquarters during which calls were made and rescinded for Vranitzky's resignation. The party was too shell-shocked to know which way to jump and so it stayed rooted to the spot; in any case no easy options seemed available. Figures published soon after the election showed a further rise in unemployment (7.2 percent) and an increase in those living in poverty, embarrassing and depressing news for a Socialist Party which had been in power for over a quarter of a century. Forceful arguments were put forward for a change from a welfare to a social state and for less bureaucratic waste and an end to state handouts to those not really in need. It was time for resolute action and long-overdue reforms, many of which had long been supported by the Freedom Party.

"Golden Oldies" of the SPÖ who had served during the Kreisky period accused the party leadership of drifting and of being unable to solve pressing problems in education and social welfare.[4] Former mayor of Vienna, Helmut Zilk, believed that it had been a mistake to oust Haider as provincial governor in Carinthia since from that time he had played the martyr role to the full. Ex-president of the Trade Union Federation, Anton Benya, noted that the secret of the success of the SPÖ had been that it could communicate with the people and regretted that it was now only Haider whose language was intelligible. Modern SPÖ functionaries could only appear on television with long-winded and complicated sentences, he said. Kreisky's old Finance Minister, Hannes Androsch, went so far as to describe Vranitzky as Haider's best "election aid" since, instead of isolating Haider in the last decade, he had isolated himself and his party. According to Benya the party had totally miscalculated the explosive potential of the immigration issue. He noted that the unions had always supported the idea of quotas and controlled immigration but had let the populist Haider have a free hand.

Haider rubbed salt into the wounds of the SPÖ. He rammed home the message that the FPÖ was the party for the workers by claiming that he had inherited the mantle of Bruno Kreisky. Vranitzky wryly retorted that he hoped Haider had paid the necessary inheritance tax, a comment appropriate to an ex-banker. In a further interview Haider compared the Freedom Party to the Palestinian Liberation Organisation which provoked the wrath of the Foreign Minister Schüssel. Haider deliberately picked the PLO once so warmly embraced by Kreisky and declared that the Freedomites were fighting to liberate the Austrian people. In an interview with *Der Spiegel* Haider referred to "pre-fascist" elements in Austria – a centralised state, a rigid division of power and party domination.[5] The FPÖ sought, he said, to free citizens from dependency on parties for jobs, subsidies and privileges.

FELDKIRCH '96

Soon after the European elections, the FPÖ held its regular party conference in Feldkirch, near the Swiss border. The mood of delegates was upbeat which was not surprising given the encouraging election results that autumn. One of the big changes announced (which already had been leaked to the press) was the appointment of a "number two" in the party hierarchy to support Haider. In parliament there had been for some time a division of labour and Ewald Stadler had relieved Haider of much of the routine business of the party in the National Council. The new number two in the party organisation was Susanne Riess-Passer who had long been at Haider's side as a loyal and diligent worker. Born in 1961 in Braunau am Inn, she had read law and business studies at Innsbruck University. She had journalistic experience and had been a member of the European Parliament as well as the second chamber in Vienna, the Federal Council. Riess-Passer was to ease the load on the leader, look after the provincial organisations and international contacts, and improve efficiency and communication. Haider's proposal to put up Riess-Passer was supported by almost 90 percent of the conference delegates. This choice was a signal that women could also make a career in the FPÖ and was to counter the ÖVP's Stenzel bonus as well as so-called power women in the other parties. The October elections had shown that the FPÖ was beginning to be attractive for women who went out to work, but Haider was aware that there were still gaps in the electoral profile of the party.

The party had been successful at national level but had yet to make a breakthrough in the small communities and towns where the village pasha still came from the old guard. There could be no question of relaxing the pressure on the governing parties and no repetition of 1995 when the party had been taken by surprise by a snap election. The party had improved its membership nationally and across most provinces (see Table 12) although this increase had been modest in comparison with the rising number of voters. The party had to work on its image, look professional and present its leader as a statesmanlike figure at home and abroad who could come up with alternatives for government.

Conference management was geared to projecting a futuristic image by means of the Internet which relayed the proceedings live to thousands of "surfers" well beyond the tiny confines of Feldkirch. A giant screen in the hall periodically flashed comments and messages of support from those who had witnessed the conference in this way. Many good wishes for the future came from the less well off and the handicapped who had been hit by government cut-backs. This large dose of pathos had the desired effect of portraying the FPÖ as the caring party and as the better workers' party with a heart for the less privileged in society. Haider in his speech made a scathing attack on the People's Party "with the Rosary in its pocket and Socialism on the brain." He had been particularly irritated by comments from leading members of the ÖVP who insisted the FPÖ was not fit to govern. The ÖVP was accused of conniving against the FPÖ to sabotage its chances of joining a political group in the European Parliament. Haider himself was confirmed as leader by 98.55 percent of conference votes.

The main document for debate related to the creation of jobs which had become a burning issue. This was linked with the motto "cut taxation" an appealing combination around which the party could easily unite.

The conference proceedings were interrupted to show a live television discussion between Haider and newspaper journalists. They could only ineffectually ask why it was that the FPÖ leader "didn't like people" to which Haider looked suitably wounded and perplexed. Then Haider revealed two major points which took viewers including his own delegates in the conference hall by surprise. He proposed the reinstatement of a new general secretary, a post abolished at the Linz conference in 1995. The post was to be filled by his press secretary the young Peter Westenthaler. Then Haider announced the intention to found a separate trade union organisation to stand up for the real interests of workers instead of functionaries.

The Austrian Trade Union Federation (ÖGB) was founded in 1945 and today consists of 14 unions and different political groups (*Fraktionen*). Recognised groups include the SPÖ, the ÖVP and the KPÖ Communists, but not the FPÖ which did not exist in the 1940s. Since the 1970s the FPÖ has had a seat on the ÖGB executive but the member is only co-opted. Like many other post-war institutions, the ÖGB had failed to move with the times and Haider decided it was high time to tackle this anomaly particularly as he was now speaking for the workers. Like the SPÖ, the ÖGB leadership had drifted apart from the rank and file and was steadily losing members. The ÖGB had not helped itself and had resisted change on many fronts such as the liberalisation of shop opening times. It had defended the principle of early retirement for rail workers even in their early 50s, when this was no longer feasible.

There had been much thinking and talking about reforming the ÖGB structure but it was Haider's television bombshell which accelerated the process. Haider sought the recognition of the *Freiheitlichen* as a political group within the ÖGB which also would bring financial accruement as well as representation in the executive bodies. Now aware that many of their own members (estimated to be around 30 percent) were FPÖ voters, the ÖGB boss, Fritz Verzetnitsch, was obliged to open discussions with Haider on a perestroika. Haider knew his success in federal elections had been impressive in the preceding decade. The bastions of social partnership were beginning to be vulnerable to his challenge. The penetration of Austria's institutions by the FPÖ had only just begun but was necessary if the party was to consolidate its gains and be involved in the decision-making process at different levels.

One area where Haider had continually attacked the reds and blacks for building up their own empires was the banking sector. This now came into the limelight and offered Haider an excellent opportunity to score points at the expense of an increasingly unconvincing coalition.

"RED BANK – BLACK BANK"

For six years Austria had been trying to privatise a bank. This project had encountered inordinate obstacles, and crucial decisions had successively been postponed. The foreign press even talked of the "jinxed bank," the Creditanstalt (CA) – "Austria's second largest bank blamed for triggering the Great Depression when it collapsed nearly 70 years ago." According to *The Financial Times*, 20 December 1996, the prolonged wrangle to sell the government's 70 percent stake in the CA "has raised serious questions about Austria's willingness to embrace a free-market economy and ditch the old system of *Proporz* where the two main parties divvy up top jobs and influence over the economy."

At the end of 1996 a new row flared up involving plans to privatise the CA which nearly wrecked the coalition. The country's largest bank, Bank Austria (BA), put in the best offer of three in a "hostile take-over" bid of the Creditanstalt. The BA offer was financially attractive especially for the Finance Minister Viktor Klima, and opened the chance of modernising an inefficient banking system while keeping the CA in Austrian hands. Another bid had come from a consortium EA Generali, led by an Italian insurer and First Austrian, the country's oldest savings bank. The third offer was less than these two and came from an Austrian retailer.

The snag with the Bank Austria option was political. The bank was regarded as a "red" socialist preserve which was making advances in an essentially "black" domain, that of the Creditanstalt. The People's Party got wind of the planned take-over late in the day and was outraged. Its leaders complained that Socialists had even informed Jörg Haider of their intentions instead of them, although this was denied by the SPÖ. Haider at any rate was pleased with the entire confusion. He greeted the BA offer whilst at the same time rejecting it in its original form. He thus neatly played off a thoroughly disorientated ÖVP against a smug SPÖ trying to pull a fast one on the coalition partner.

The coalition pact agreed between SPÖ and ÖVP on 11 March 1996 supported further privatisation. It expressly favoured a sell off of the federal stake in the Bank Austria and Creditanstalt with the proviso that Austrian interests and improvements in the economic substance of the firms should be taken into account. Both the SPÖ and the ÖVP now accused the other of breaking this pact. The People's Party complained that the BA option was not a genuine privatisation scheme but "communalisation" or quasi-nationalisation by the back door. It pointed to the close links between the bank and the red-controlled Vienna town hall. A CA/BA merger would create a bank ranking about 30th in size in Europe and with a 25 percent share of the domestic market. Through this Bank Austria would scoop a near monopoly in fields such as domestic investment banking and export finance. The SPÖ Finance Minister, Viktor Klima, stuck to the argument that the government sell off would be the best bet for tax-payers and for Austria's economic future.

Just before Christmas 1996, the ÖVP saw no alternative but to join forces with the FPÖ in parliament in passing a resolution which called on the government to implement a full privatisation scheme for both the BA as well as the CA. The ÖVP/FPÖ alliance was in operation a week later in the Federal Council, the *Bundesrat*, against the votes of the SPÖ. The Finance Minister insisted he had to accept the best offer and was not legally bound by such a parliamentary resolution. The coalition just one year after the premature elections of 1995 had once more run aground. The same arguments and language could be heard which had a year before precipitated a sudden winter election. The ÖVP felt cheated and accused the SPÖ of breaking the coalition agreement whilst the SPÖ rounded on the People's Party for shady double-dealing. Much was at stake. The ÖVP had steadily been losing influence in the banking sector and its claims to be a party of business and for the economy were beginning to look wobbly. The SPÖ had realised politically it had its back to the wall but that economically there was still all to play for to shore up its power base.

A bemused American newspaper described the dilemma as "red bank – black bank." The ÖVP feared that if the BA offer was taken up it would be "red bank – red bank." The FPÖ could not have wished for any better confirmation of the thesis that the reds and the blacks had carved up Austria from top to bottom in all important spheres. Even more welcome was the fact that the confirmation was provided by the "predators" themselves.

Christmas came and time was running out for the ÖVP. It set the wheels in motion for a special sitting of parliament in the new year to thwart the Socialists' banking ambitions. In preparing for this, the ÖVP was obliged once again to go to the Freedom Party for further support in parliament; in doing this it risked destroying the coalition – a hazardous tactic which could not be toyed with every year. A sick President Klestil appealed to the coalition parties to come to their senses and reach a solution without new elections. Unfortunately it seemed that a dialogue between the chancellor and his vice-chancellor Schüssel had totally broken down. The leader of the ÖVP in parliament, Andreas Khol, daily beat the drum against the Socialists and defied the finance minister to take a decision on the takeover before the special sitting of the National Council. Minister Klima seemed concerned but determined not to break the law or to be bullied. He had already delayed the sale once, reopening the tender to allow bidders to improve on their original offers and a final decision was now imperative to avoid Austria becoming the laughing stock of the financial world. The count-down had begun for the special parliamentary sitting in mid-January and an FPÖ/ÖVP document had been drawn up and even signed which agreed on a common strategy in the debate. It seemed the two coalition partners were slithering towards early elections which no one could really want with the exception, perhaps, of Jörg Haider.

At this point a coalition committee was invoked which was laid down in the coalition pact to sort out serious cases of disagreement. A marathon crisis session comprising top politicians and bankers from both sides of the coalition met into the early hours of the morning to hammer out a deal. Finally it was agreed that Bank Austria's offer for control of the Creditanstalt would be accepted but only if certain conditions were met. According to this, the Vienna municipality's share in Bank Austria is to be reduced to less than 25 percent within five years and later cut further to less than 20 percent. The employees of CA will be able to buy shares and their jobs should be guaranteed. The Creditanstalt should remain intact as a legal entity for five years:

Coalition Agreement Concerning the Sale of the CA

1. Within five years the voting rights of the AVZ and the Wiener Holding should be reduced to below 25 percent. In another two years the percentage must be reduced to below 20 percent. Until then the voting rights for shares above this level will be exercised by trustees. If within these periods (5 or 7 years) the stated percentages cannot be reached, trustees will also be appointed.
2. The 19 percent state-owned stake in the Bank Austria must be privatised this year.
3. A guarantee is given that the CA will remain an independent company for the next five years.
4. Moreover a guarantee is given that the jobs in the CA will be preserved.
5. Investkredit and the control bank must surrender their CA shares minus the GiroCredit shares to other shareholders. The AVZ must offer all its GiroCredit shares as soon as possible. Moreover the Bank Austria may not participate in the privatisation of the PSK.
6. The Bank Austria may not strip the assets of the CA and sell them.
7. All financial institutes have to render a "liability compensation" on the basis of an opinion drawn up by international experts if they do not renounce municipal liability.
8. CA shareholders will be offered a swap for their shares. Moreover ordinary shares may be compensated for up to a maximum of 200,000 Schillings in cash.
9. CA employees may purchase shares up to a maximum value of 500 million Schillings.
10. The Privatisation Act requires the agreement of the Federal Government for the sale of Federal assets.
11. The term of the exemption provisions for consolidation expires on 31 December 1998.
12. A take-over law will be drawn up.
13. The Bank Act will be amended in the areas of "Renunciation of Liability and Liability Compensation."
14. The Vienna Stock Exchange shall be transformed into a public limited company by the summer 1997.
15. Points 10-14 are to be passed by summer 1997 by means of a Government Bill.

16. In 1997, 1998, and 1999, one billion Schillings per annum from the sales proceeds will be made available for research and development and export promotion.
17. If the AVZ fails to vote in the sense of Point 1 within four weeks, the CA sale will be reversed.

According to *The Financial Times*, 13 January 1997, the solution was an "Alpine fudge":

> Postponing proper rationalisation and privatisation of the sector for five years or more will drag out a process that has already been delayed. That said, the solution could have been worse. At one point it looked as if even less rationalisation would be permitted and that Bank Austria would be controlled by Vienna indefinitely. So one should be thankful for small mercies.

The paper also attacked the poor offer for minority shareholders of the Creditanstalt, a point reiterated by the Freedom Party. No one was really convinced that the pious words on safeguarding jobs would really stand the test of time.

The deal was generally interpreted as a coup for Finance Minister Klima while the ÖVP had tried to salvage what it could from its scorched earth policy. The agreement left the ÖVP with the embarrassment of the special sitting of parliament in which it had planned to join forces with the FPÖ against the Socialists. The sitting could not be called off and the ÖVP was obliged to ditch the FPÖ and return to the coalition fold during the debates. The Freedom Party screamed "traitors" and scorned the ÖVP for permanently capitulating and reneging on its word. On 9 January, the ÖVP parliamentary party had agreed to a "political consensus independent of the decision of Minister Klima" with the FPÖ. A meeting was scheduled for 13 January, a day before the special sitting of the National Council, between the People's Party and the Freedom Party in the parliamentary rooms of the ÖVP. When the FPÖ delegation showed up there was no one else there. Some confusion and embarrassment followed during which the FPÖ was hastily invited over to the ÖVP-run Ministry of Economics only to be shown the back-door to avoid a television crew. The SPÖ, too, ridiculed the zig-zag policy of the ÖVP and looked suspiciously at the return of its bruised partner.

The FPÖ position on the privatisation of the CA was documented as follows:

1. No detrimental effect for the tax payer – sale to the highest bidder. According to the Act passed with the votes of the SPÖ and the ÖVP authorising the sale of the shares in the CA-BV, the Minster of Finance, as the owner, is obliged to sell the shares to the highest bidder while taking into account the national interest.
2. No retrospective changes to the law.
3. An international group of experts to monitor CA sale.
4. Privatisation of the Bank Austria (BA) as the next step.
5. Privatisation Act with an amendment to the Bank Act. The 48 municipal banks, which are in a similar position to the Bank Austria, and the City of Vienna should be completely privatised within the next two years. If there is a danger that this schedule would lead to a price loss, privatisation should be carried out through a trust agency within a maximum period of five years.
6. Liability clause for the duration of communal liability.
7. Improved position for small shareholders. It is the international practice that if an institutional investor acquires a certain percentage of the stock capital (approx. 33 percent) it must also be willing to make a take-over bid to the minority shareholders on the same conditions.
8. New evaluation of the stake in the Bank Austria owned by the state (18 percent) in the event that the state-owned shares are sold to the Bank Austria.
9. Debate on the consolidation privilege enjoyed by the Bank Austria in the special sitting of parliament. The exception that was granted should be shortened to the year 2000.
10. Changes in liability can only be made "pro futuro."
11. Exclusion of liability for the state.
12. Solution for the control bank. If the BA takes over the CA at least 60 percent would be under the same ownership.
13. Reduction of the concentration of power. It is intolerable for the management of Bank Austria to acquire the management of the AVZ while the General Director chooses his own members of the Supervisory Board.
14. Cooperative reform. What applies to the bank sector must also apply to the cooperative sector.

Vranitzky "Adieu"

The fallout from the Creditanstalt fusion with the Bank Austria enhanced the stature of Finance Minister Klima whose popularity soared in the country. A rather jaded Chancellor Vranitzky had taken a back seat during the vital negotiations. Shortly after the whole affair had been settled Vranitzky suddenly announced he would step down as chancellor and SPÖ leader in favour of the "crown prince" Klima. Vranitzky had been chancellor for over 10 years but had come under increasing fire from his own people for his aloofness and irresolution. He had kept the Socialists in power for over a decade, but now seemed unable to inspire even himself.

An era had come to an end. An era which had begun with the election of Jörg Haider as FPÖ leader in September 1986 and which had witnessed bitter political duels. Haider after many rounds with Vranitzky had finally won on points. The policy of marginalisation was of dubious value as a long-term strategy, and Vranitzky bid "adieu."[6] Haider had long predicted Vranitzky's departure and had talked of a new relationship with the SPÖ if Klima should come to power. Chancellor Klima promptly ruled out any liaison with the "Haider FPÖ" but most sensed this option would be opened up if the time was ripe. By now it was an open secret that all three parties were engaged in talks about talks with each other. Any possible permutation was possible in the future.

The FPÖ had a new potential suitor and the coalition had a new leader, an affable type with a toothpaste smile and a liking for first names. Klima had experience in marketing and commerce. In his first major policy statement to parliament, Klima spoke of the *Angst* which had gripped the country. He recorded that no hard and fast guarantees could be given to those who looked for a job for life. The speech had no uplifting phrase and contained no promises or big solutions. The new chancellor talked of the need for change and of helping people out of the poverty trap. There was to be less bureaucracy and more deregulation and a balance between the state and the market. The *Angst* remained: Klima represented for Haider a new challenge and a new opportunity.

The banking scene continued to dog relations between the coalition partners damaging both the ÖVP and the SPÖ under Klima. Just after Easter 1997 the sensational suicide of a leading banker and Social Democrat, Gerhard Praschak, shocked the public and exposed party political interference in the banking world. Praschak, a former economic adviser to Vranitzky, shot himself in his office after dispatching by post documents to the press and the opposition parties which allegedly showed that he had been a victim of political intrigue. According to Praschak, he had been sidelined for a top job to make way for Rudolf Scholten, former Socialist Minister for Transport and the Arts. Praschak's files implicated Chancellor Klima, the Finance Minister and the Mayor of Vienna, all Social Democrats, in these alleged sordid backroom deals. In his suicide note, Praschak further accused the Bank Austria, the main shareholder in his bank, of a tax evasion scheme through a covert profit pay-out. All his accusations were denied by those concerned. Praschak was one of the best paid top managers in the country and his desperate act bewildered many Socialist members.

Jörg Haider jumped on the case to accuse Chancellor Klima of deliberately forcing Praschak out of the bank for political reasons. He compared the incident to the connivings of the Italian Mafia but added that whereas in Italy the Mafia was private, in Austria it inevitably was state run. The Freedom Party called for a parliamentary commission of inquiry to investigate the Praschak affair. The People's Party, although it welcomed the opportunity to embarrass the SPÖ, preferred to argue for less party politics in the banking sector. It shrank from starting off a lengthy process of investigation in parliament which could only provide a stage for the opposition parties. The ÖVP also feared that such commissions could become too fashionable and it was resisting pressure from the Greens for a parliamentary inquiry into the murder of Kurdish opposition leaders in Vienna in 1989 carried out by an Iranian hit squad. Former ministers and politicians from the People's Party had been accused of a cover-up (after yielding to blackmail from the Iranian government) allowing the suspects to flee Austria.

All three opposition parties were outraged with the repeated rejection of the coalition to agree to parliamentary committees of inquiry. This frustration led to a curious spectacle when the Greens, the Liberals and the Freedom Party joined forces in a temporary boycott of all parliamentary committee work as a protest against the obduracy of the government.

Soon after Praschak's suicide, ghosts of the Bank Austria deal returned to haunt the troubled coalition. A prospectus of the bank was discovered which had been published in America and revealed that a "Red" German bank had a contract guaranteeing it the right to first refusal on the planned shares of the almost half of the Bank Austria. The revelation that Germany's third largest bank could take over a majority interest in the BA staggered the ÖVP who found it hard to believe that leading Socialists, including the former Finance Minister and current Chancellor Klima, had not been in the know. Leading ÖVP politicians accused the coalition partner of deceit and treachery, making it clear that it would stick to the coalition in the interests of Austria but for no other reason. The SPÖ tried to pass the issue off as a tempest in a teacup, maintaining that such agreements were completely normal. Whatever the interpretation, it was clear that the ÖVP had once again been duped by the Socialists who were better at playing banking politics. Both partners in the coalition it seemed were working for Haider, who departed for a six week course in economics at Harvard University.

Notes

1. See J. Haider, *Friede durch Sicherheit* (Vienna, 1996) and *Freiheitliche Europapolitik*, 5/96, both published by the Freiheitliche Akademie.
2. Interview with Sichrovsky. November, 1996, Feldkirch party conference. After his election Sichrovsky was criticised for spending most of his time in Chicago rather than Strasbourg.
3. Interview with Firlinger, January 1997. His analysis was born out by comments of further defectors from the Schmidt party.
4. See interviews with Karl Blecha, Helmut Zilk, Anton Benya, Hans Mayr and Hannes Androsch in *News*, 47/96.
5. *Der Spiegel*, 43/96.
6. An indictment of the Vranitzky period with respect to Haider can be found in H. Czernin, *Der Haider-Macher. Franz Vranitzky und das Ende der alten Republik* (Vienna: Ibera & Molden, 1997).

TABLE 10
FINAL RESULT OF THE EUROPEAN PARLIAMENT ELECTION, 13 OCTOBER 1996

A total of 3,794,145 valid votes was cast, and the election turnout was 67.73%. The votes were distributed between the parties standing for election as follows:

	% Votes	Votes
Austrian People's Party (ÖVP)	29.65	1,124,921
Austrian Social Democratic Party (SPÖ)	29.15	1,105,910
Austrian Freedom Party (FPÖ)	27.53	1,044,064
The Greens	6.81	258,250
Liberal Forum (Libs)	4.26	161,583
Die Neutralen	1.28	48,600
Forum Handicap	0.86	32,621
Austrian Communist Party (KPÖ)	0.47	17,656

The 21 seats in the European Parliament were thus allocated as follows:

ÖVP: 7; SPÖ: 6; FPÖ: 6; Greens: 1; Libs: 1.

The previous distribution of the 21 Austrian seats in the European Parliament was based proportionally on the outcome of the last parliamentary election of 17 December 1995:

ÖVP: 6; SPÖ: 8; FPÖ: 5; Greens: 1; Libs: 1.

Source: Ministry of Interior, Vienna

TABLE 11

THE 1996 EUROPEAN PARLIAMENT ELECTIONS: A SOCIO-DEMOGRAPHIC ANALYSIS

Voter Group		Share of Voter Group (%)				
		SPÖ	ÖVP	FPÖ	Libs	Greens
General Results		29.2	29.7	27.5	4.3	6.8
Men	Total	26	28	32	6	7
	Working	22	29	35	5	8
	Retired	38	26	31	1	1
Women	Total	32	31	25	4	6
	Working	25	29	33	4	7
	Housewives	38	38	13	4	2
Voters aged:	Under 30	21	25	21	17	12
	Between 30-44	25	25	35	4	8
	Between 45-59	25	36	30	3	5
Voters over 60	Total	43	29	22	1	2
	Male Pensioners	38	26	31	1	1
	Female Pensioners	44	31	17	1	4
Blue-collar workers		24	21	50	1	2
White-collar workers		26	28	30	5	9
Voters who had swung from another party in previous nation-wide elections		10	26	40	8	14
Non-voters (political sympathies as expressed by non-voters questioned)		23	13	15	5	4

Source: Fessel +GfK: post-election telephone survey of 1150 individuals.

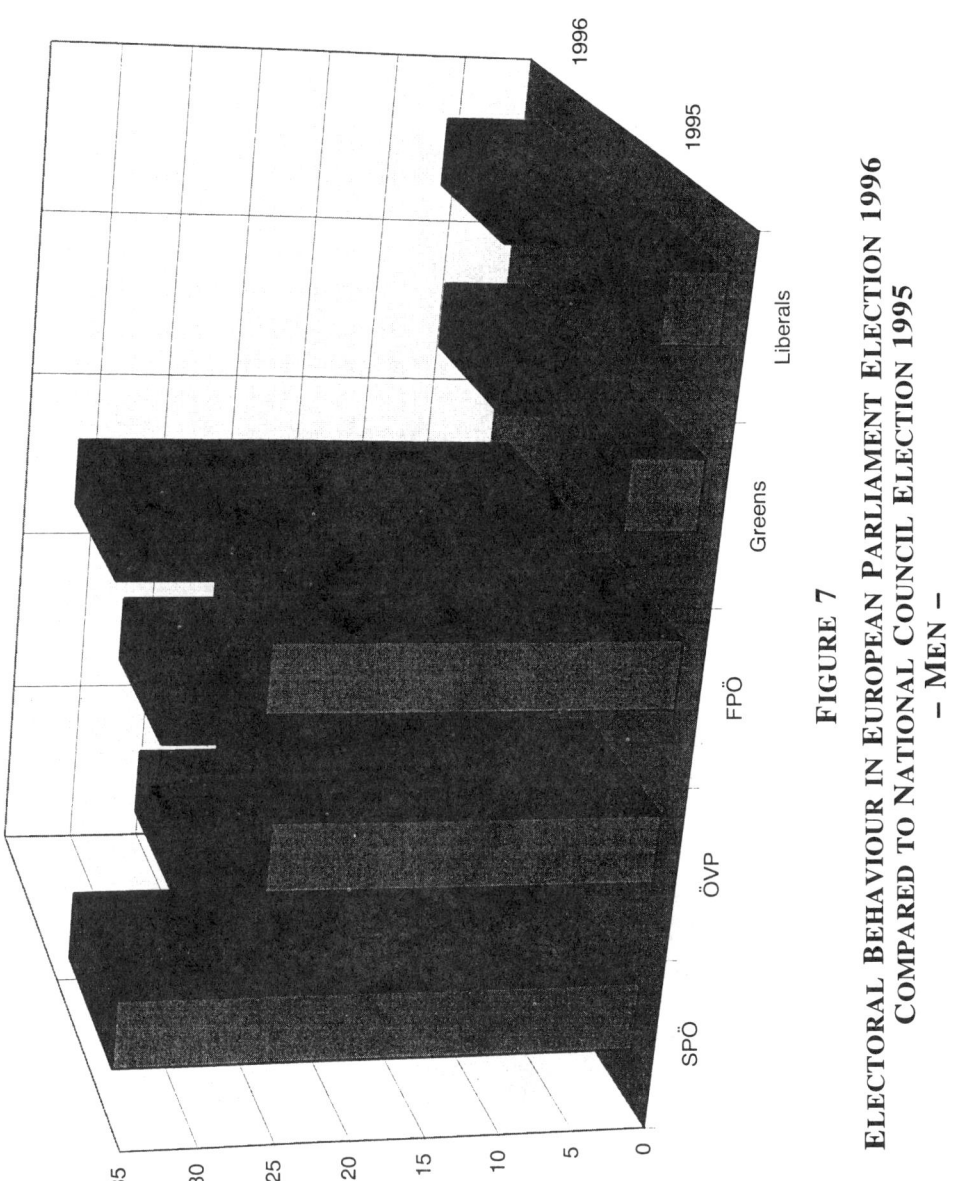

FIGURE 7
ELECTORAL BEHAVIOUR IN EUROPEAN PARLIAMENT ELECTION 1996 COMPARED TO NATIONAL COUNCIL ELECTION 1995
– MEN –

Source: *Kurier*, 15.10.1996 (drawn on research presented by F. Plasser and P. Ulram).

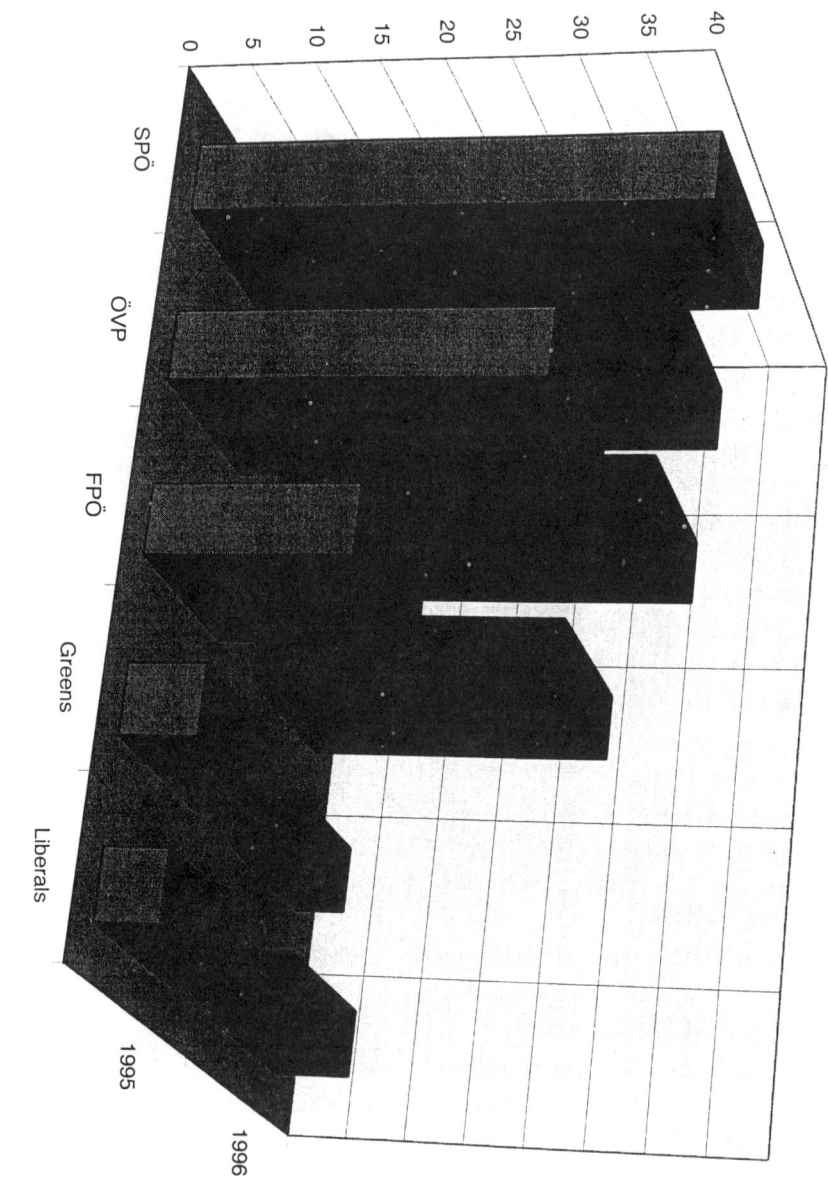

FIGURE 8
ELECTORAL BEHAVIOUR IN EUROPEAN PARLIAMENT ELECTION 1996 COMPARED TO NATIONAL COUNCIL ELECTION 1995
– WOMEN –

Source: *Kurier*, 15.10.1996 (drawn on research presented by F. Plasser and P. Ulram).

EURO ELECTIONS '96 AND AFTERMATH 193

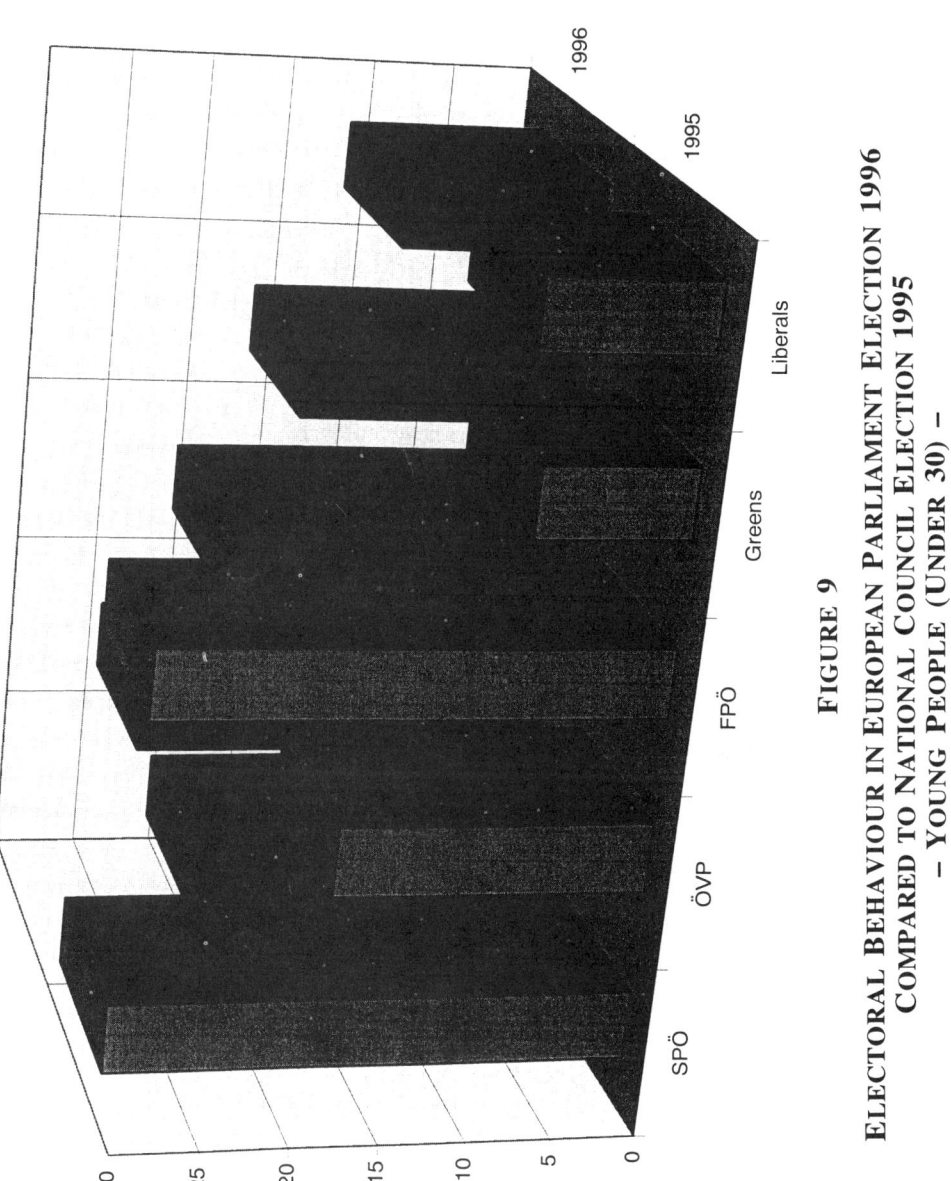

FIGURE 9
ELECTORAL BEHAVIOUR IN EUROPEAN PARLIAMENT ELECTION 1996
COMPARED TO NATIONAL COUNCIL ELECTION 1995
– YOUNG PEOPLE (UNDER 30) –

Source: *Kurier*, 15.10.1996 (drawn on research presented by F. Plasser and P. Ulram).

194 THE HAIDER PHENOMENON

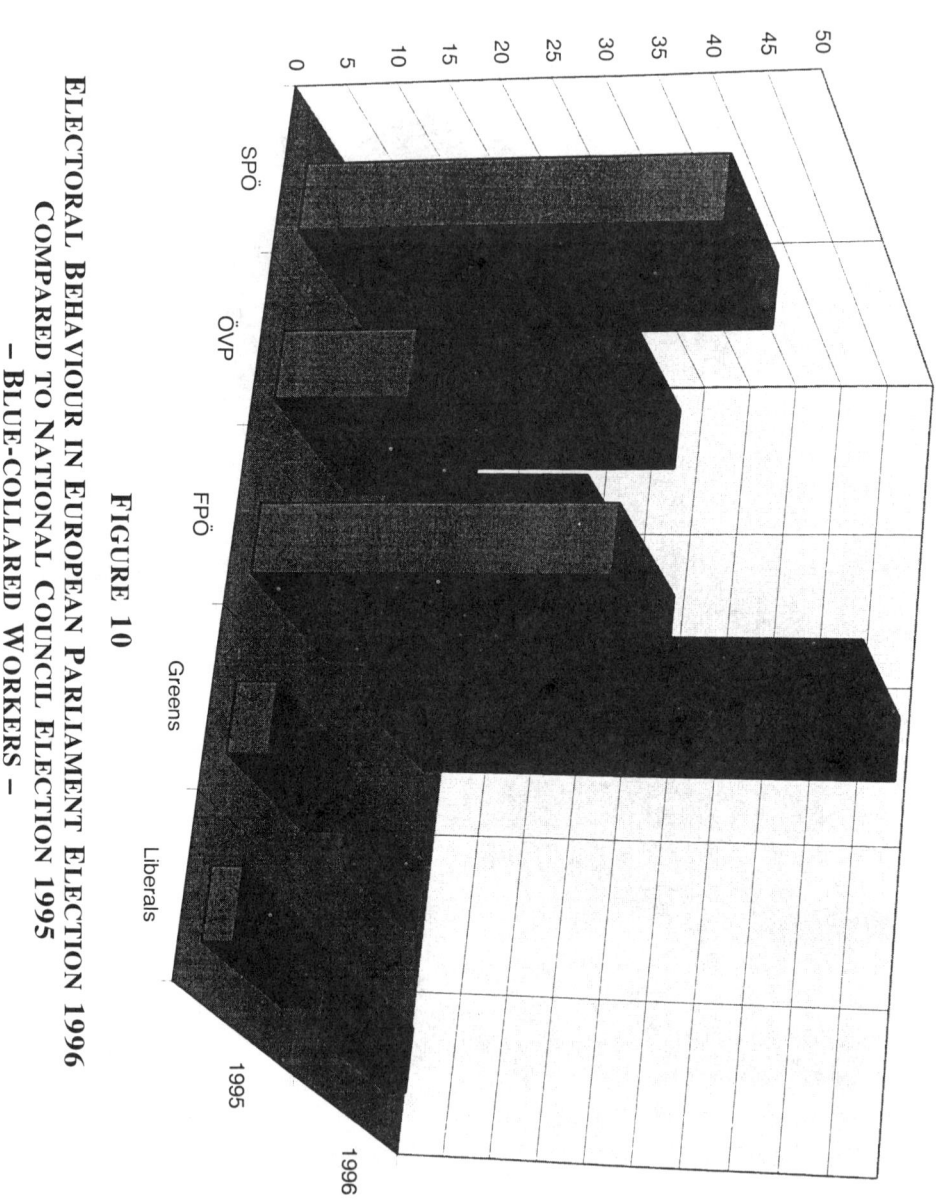

FIGURE 10
ELECTORAL BEHAVIOUR IN EUROPEAN PARLIAMENT ELECTION 1996 COMPARED TO NATIONAL COUNCIL ELECTION 1995 – BLUE-COLLARED WORKERS –

Source: *Kurier*, 15.10.1996 (drawn on research presented by F. Plasser and P. Ulram)

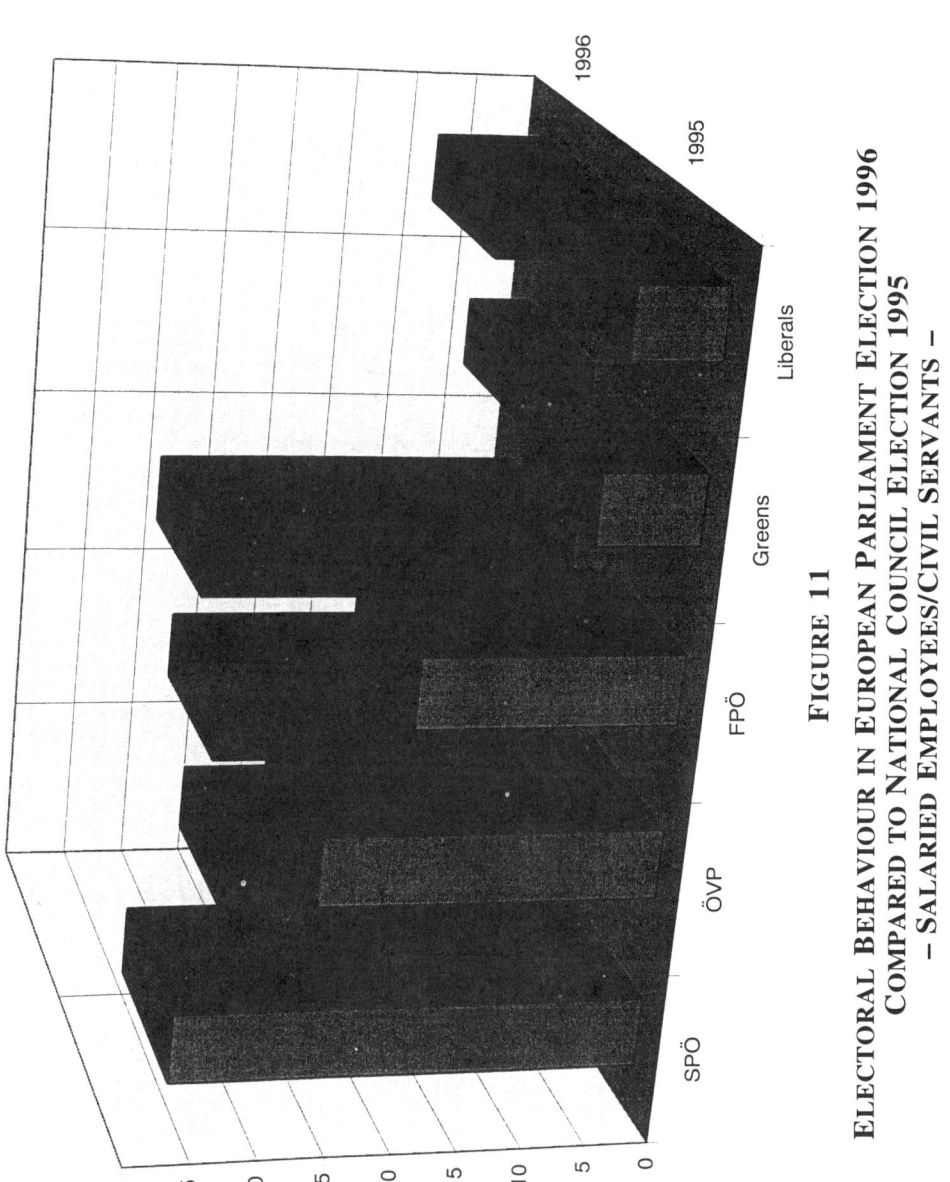

FIGURE 11
ELECTORAL BEHAVIOUR IN EUROPEAN PARLIAMENT ELECTION 1996 COMPARED TO NATIONAL COUNCIL ELECTION 1995
– SALARIED EMPLOYEES/CIVIL SERVANTS –

Source: *Kurier*, 15.10.1996 (drawn on research presented by F. Plasser and P. Ulram).

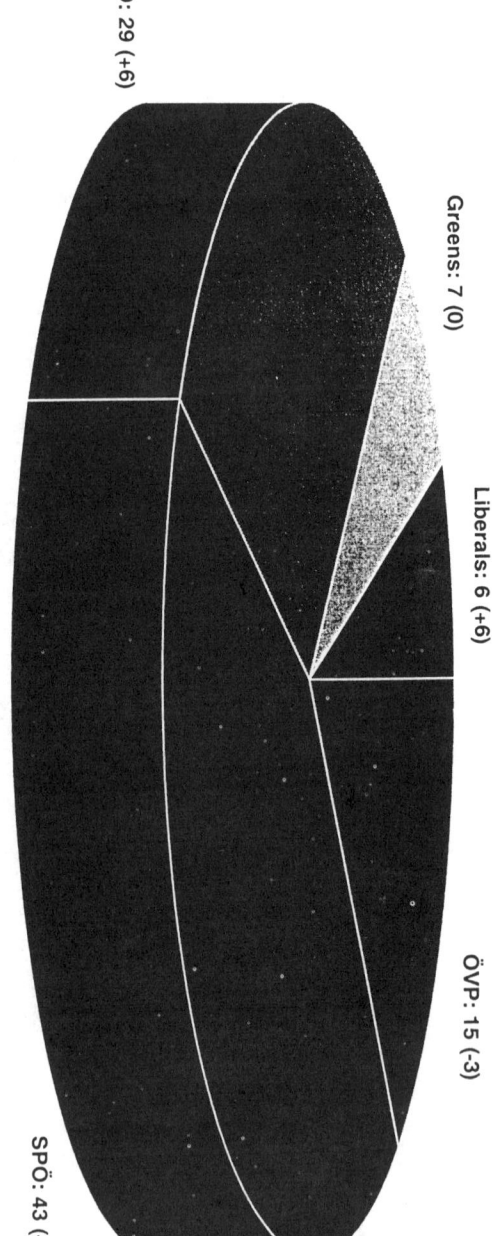

FIGURE 12
SEATS IN VIENNA MUNICIPALITY, 1996

Source: Vienna Town Hall.

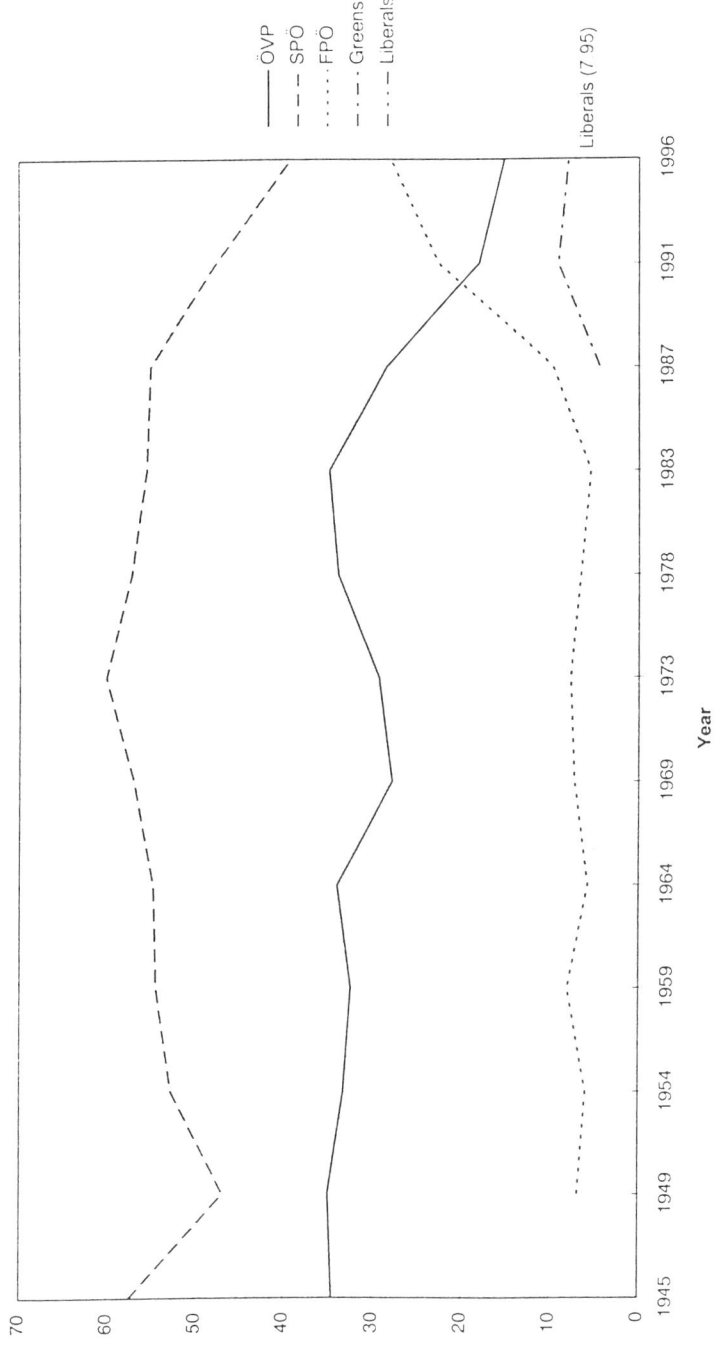

FIGURE 13
MUNICIPAL ELECTIONS IN VIENNA, 1945-1996

Source: Vienna Town Hall.

TABLE 12
FPÖ MEMBERSHIP 1992-1996

Members in Region	1992	1994	1996
Burgenland	1,015	1,093	1,027
Carinthia	6,904	7,206	6,543
Lower Austria	4,642	5,398	5,402
Salzburg	4,061	4,138	4,508
Styria	5,325	5,657	6,246
Tyrol	3,443	3,568	3,716
Upper Austria	10,771	11,099	11,667
Vienna	2,388	2,611	3,502
Vorarlberg	2,713	2,994	2,233
Austria	41,262	43,764	44,844
Austria - Organisational Density (Ratio of Voters at Previous Election, in Parentheses, as %)	5.27 (1990 Election)	4.20 (1994 Election)	4.23 (1995 Election)

Sources: FPÖ Report to Conference 1995 and 1996, and Ministry of Interior, Vienna

13

THE ENTERTAINER

Haider has fascinated many journalists, politicians and intellectuals who have seemed mesmerised by his presence. Countless television round table discussions and interviews invariably ended up with a debate on Haider, even when the original topic was something else. Haider's "presence" could be felt even if he was not in the studio. The Freedom Party leader dominated the political agenda and his statements demanded a response. Vranitzky spent much of his time trying to make sure that Haider would not be his successor as chancellor. This he did but he could not stem the Freedomite advance and many criticised Vranitzky's disastrous policy of exclusion with respect to the FPÖ.

The writer Robert Menasse, no supporter of Haider, came to the following conclusion, "Haider criticises many things which the Left always criticised. I can't understand why suddenly intellectuals say that everything they always criticised, is now all right just because a political opponent also suddenly criticises it."[1] The only difference, according to Menasse, is that whereas the intellectuals were ineffectual, Haider has had an impact.[2] These people had been complaining about social partnership and historical myths of the republic for years. Suddenly they turned hysterically on Haider for uttering the same sentiments with the accusation that he sought to destroy the system. The left intellectual critique was essentially closed and sterile; it preached to the converted and changed nothing.

When, for Menasse, a transformation process started in Austria after 1986, it came not as a result of the intelligentsia but caught them by surprise. Suddenly more people in Austria were beginning to question what the intellectuals for years had been saying in the wilderness. A greater scepticism was apparent which challenged historical taboos and social partnership. Suddenly left intellectuals were confronted with what they had so long yearned for – an opposition which sought to move a stultified society in the direction of a modern democracy. But, concluded Menasse, "the Austrian intellectuals were not euphoric, not even encouraged but they were appalled and even hysterical." These people detected an ominous shift to the right when all that had happened was that the mood in the country had changed in favour of more openness and opposition. They read "Fascism" for the F movement and overlooked the importance of two other parties relatively new in parliament, such as the Greens and the breakaway Liberals. "They saw the danger of the destruction of the Second Republic, and they could not think of anything else to defend the republic other than to suddenly defend the myths, the legends and the taboos which they themselves had always criticised, and they do this because they think that these myths, legends and taboos belong inseparably to the republic which now must be defended - instead of seeing that they thus help to manufacture just the atmosphere in which Jörg Haider cleverly can profit." Many writers and artists on the left were seriously worried about Haider's rise and feared that life would be made difficult if he were in power. Haider talked of the "state artists" who were anxious of losing their generous subsidies. Such artists dramatically talked of going into exile should Haider come to power.

For Menasse, Haider fulfilled a logical function in a country undergoing a normalisation process to an ordinary western democracy with a developed parliamentary system, alternative governments, less social partnership and an effective opposition. For him it was predictable that the opposition should come from the right after so long a period of social democratic rule. It was even more logical that this would not come from the moderate right since its representative, the ÖVP, was in coalition with the Social Democrats. Not only that but the People's Party was one of the main pillars of social partnership. Haider had contributed to the democratisation process in Austria and for Menasse Austria was a more lively democracy than ten years ago. He hoped that the Greens and Liberal Forum would cease to look at Haider so hysterically and contribute to this positive process from the left of the political spectrum.

Many writers and political observers criticise what they see as the ossified, sclerotic institutions in Austria. They doubted whether very much had really changed over the last years despite Haider's advance. Two-party dominance may be on the wane in federal elections but the party book – red or black depending on geographical location – is still useful for a job, promotion and an easier life. Critics of Austrian politics confess to never having witnessed a real change in the power set-up since the war. Moreover at lower levels of administration and government in the villages and towns of Austria, life goes on much as it always did. Around 70 percent of mayors are from the People's Party (compared with about 1.5 percent for the FPÖ) and provincial governors in seven of the nine provinces (*Länder*) come from the ÖVP; only Burgenland and Vienna have Red provincial governors.

There is much disbelief at the resilience of Austrian political specialties even by the natives themselves aware of the fact that change has occurred almost everywhere else in contemporary Europe. Changes at the top do not always filter down to the layers below which can remain rigid and obstinate. Even these critics believe that Austria cannot remain immune to change for ever. Sooner or later they feel something will happen but if later then the pace of change could be too much for the democratic system to contain. This lack of confidence in the ability of the democratic republic to deal with big changes has influenced people's attitude to Haider. Many are convinced that Haider could be dangerous. He is too radical and too bent on change and for many seeking change from the wrong quarter.

Haider's project to change Austria is built on a potpourri of ideas from the right and the left, from neo-liberalist economics, conservative slogans and old socialist rhetoric. It is radical and also conservative. For over 25 years the Social Democrats have built up spheres of influence in economics, culture and politics. Such a long spell in office, whatever the party denomination, can lead to a deadening of democracy and an insensibility to people's needs. Logically in a country where socialist thought is so deeply ingrained, to introduce conservative values requires firstly a radical approach. Even with a million voters behind him, Haider must at some stage find a partner, given the nature of the electoral system. Paradoxically the quest for a non-Socialist Austria may only be realised through a partnership with the SPÖ. "Socialism" in Austria is not merely resident in a specific party, it is more a way of life. Its tentacles reach out to groups in the People's Party where innovation and entrepreneurship are alien. Its spirit pervades the traditional clientele of the ÖVP where the yearning for protection and security has deep roots. The industrial structure of Austria – predominantly small and medium sized firms – has further nursed this desire. Austria's imperial past, the reverence for authority and quasi-medieval guilds have also reinforced resistance to modernisation. The Christian Social origins of the ÖVP give it a conscience which pulls it back from contemplating anything remotely comparable to a Thatcherite course. The Socialist Party of Austria, on the other hand, is pragmatic and driven by a thirst for power. Under Vranitzky the

SPÖ became one of the most conservative parties in Austria. It sought to keep more or less everything the same, from neutrality to the Concordat with Rome. At the same time it replaced social competence with economic expertise and banking know-how. The ÖVP lost economic competence but failed to find a replacement.

Most people in Austria officially grumble that nothing much changes in the country, but deep down are suspicious of anything new or radical. In 1995 Haider promised renewal and change and for the first time suffered electoral losses especially in eastern Austria where the passion for the status quo is the greatest. If Haider can convince the electorate he would really keep good old Austria, then his chances at the polls could be greatly enhanced. In this way everything could carry on much the same even with Haider as chancellor.

Haider was fond of saying that his was the only opposition party which was in effect in government if not in office. The FPÖ set the tone whether on privileges for politicians or the Federal Reserve Bank. These issues were then discussed and occasionally somebody resigned one post or the other. But maybe more would have changed without Haider. His personality alone cemented resistance to change and united political opponents in a common cause which was built around the theme "Stop Haider."

One real problem for Haider and the FPÖ was the brown connection. To continually push Haider and his party into this corner however was a simplification and contained the danger that it could become a self-fulfilling prophesy. Haider had no possible electoral interest in spontaneously appearing at Krumpendorf. This was not an act of populism since most there were not able to vote in Austria or were so old they could not guarantee the future of the FPÖ. Haider perhaps felt the need to show that he would not be bullied by official Austria and would turn up to events and talk to people of his choosing irrespective of the consequences. In terms of political tactics this was not smart but perhaps a reaction to the incessant portrayal of Haider in the media as the "F *Führer*" of Nazi parents.

In September 1996 Haider celebrated 10 hard years as leader of a party he had radically changed. The FPÖ had become a menace to the established parties and an electoral force to be reckoned with throughout the country. In Europe it was ostracised but recognised as one of the most successful right radical parties. Haider had won over workers, wooed industrialists, had spoken to veterans of the Waffen SS and had made a bid for the Jewish vote.

Haider's greatest talents were his versatility, his powers of persuasion and his ability to communicate and entertain. The Haider phenomenon relied on rhetorical genius combined with a flair for sensing what was rotten in the country. None of these attributes could be said to be especially undemocratic, but many saw in this a fanatical pursuit to get into power whatever the cost. On the other side the almost obsessional neurosis of some to stop Haider, come what may, backfired.

The aftermath of the European parliamentary elections in 1996 showed that reforms were more pressing than ever. The columnist Peter Rabl writing in *Kurier* referred to the "crisis of the Second Republic" and the exhaustion of its political representatives and structures. He predicted the implosion of the system along Italian lines with the internal collapse of the party-run social partnership model. The leader of the ÖVP in parliament, Andreas Khol, imaginatively talked of the need for the coalition to be "turbo-charged" to provide the impetus to break through the backlog of overdue reforms. The trenches have been dug between Jörg Haider and "the rest" in Austrian politics. A gulf is also opening between outmoded structures and institutions and those forces recognising the need for liberalisation and modernisation. Jörg Haider can claim to have long championed the cause to weed out medievalist-style corporatism in Austria. The obduracy of the chamber state to move with the times has provided the unconventional populist with electoral success. Reforms that have been introduced failed to attack the problem at the roots and whetted the appetite of the public for further measures. Austria now faces many external challenges including the globalisation of the economy and increased competition.[3] These cannot be solved with hackneyed formulas and bankrupt ideologies. Parties have become alien from the people. The idea of a political party as a kind of spiritual home which provided help in times of need no longer works. The main parties in Austria have not just lost votes but confidence. There is now a crisis of confidence and a lack of trust in the governing parties to come straight and solve problems.

The current political elites have been plagued by *Angst*, centred on fear of Haider and fear of the consequences of change. In the 1990s Haider has enjoyed a fool's license. The loss of the governorship of Carinthia gave him the freedom to criticise without responsibility. His opponents thought they had removed a political enemy but were confronted instead with a loose cannon they could not control. Operation "Stop Haider" not only failed but propelled his movement into an increasingly dangerous position for the governing elites. This is not irreversible since Haider has periodically shown a capacity to stop himself when all others have missed the mark. His own worst enemy, Haider has often outmaneuvered, and then reinvented, himself. His talk of renewal and change alarmed not only his enemies but also his supporters.

Despite this Austria is going through a painful process of modernisation. A broadcasting reform is on the way and more flexibility in working hours is being introduced. There is still a tail-back of problems in education, health and pensions which need urgent attention, but there are signs that this urgency is at least being recognised, a ponderous step in the right direction. The need for change was often acknowledged by Haider's enemies but unaccountably they seemed paralysed to act on their own diagnosis. In 1992 Vranitzky spoke of the need to rid the Austrian system of its guild-like restrictions; he waxed lyrical on deregulation and a "democratisation offensive" not to mention a "modern and open society."[4] This was echoed by the political journalist Hans-Henning Scharsach, one of Haider's most scathing critics: "Voters do not expect recipes against the FPÖ but recipes for the country. Whoever brings Austria forward will leave Haider behind."[5] This was to be the prescription to stop Haider but where was the medicine? If such reforms would at long last be implemented they could ultimately render the Haider factor superfluous. If not then the ground swell of discontent will surge and then Haider's party could find itself in power. Haider is backed not only by malcontents and déclassé elements but also by many idealists who dream of a new dawn should he come to power. Many of his fans believe he could magically transform the country but they could be bitterly disappointed if he failed to deliver the goods promptly. Much like the Social Democrats in the First Republic, Haider's FPÖ has been shut out from the main power house of the republic. Like the Social Democrats in the 1920s, Haider has engaged in verbal radicalism but if given the chance could opt for moderate policies in practice. His followers who take his radicalism at face value could be quickly disillusioned. This in the long run could be the real danger to Jörg Haider and his party. It would however also pose a threat to democracy and stability in Austria. Haider has acted as a kind of safety-valve to date by providing an alternative in the parliamentary arena to those fed up with the governing elites. If in turn the new god-father fails, more drastic solutions will be sought in desperation by those who feel cheated and once more let down.

NOTES

1. Quoted in *Die Presse*, 6 October, 1996.
2. Quoted in *Falter*, 41 and 46/95. See also R. Menasse, *Die sozialpartnerschaftliche Ästhetik* (Vienna: Sonderzahl, 1996).
3. According to a report produced by the Swiss Institute for Management Development (March, 1997), the Austrian economy has become relatively less competitive in the period 1994-7. In 1994/5, Austria ranked as number 11 in the league table of the most competitive countries in the world. In 1995/6, it had dropped to sixteenth place but by 1996/7, fell behind Malaysia to nineteenth place. For years the top places have been held by the USA, Singapore and Hong Kong.
4. Quoted in A. Thurnher, *Franz Vranitzky* (Frankfurt: Eichborn, 1992), p. 43.
5. H. Scharsach, *Haiders Kampf* (Vienna: Orac, 1992), p. 236.

INTERVIEWS

INTERVIEW WITH KURT WALDHEIM
VIENNA, DECEMBER 1995

Austrian Foreign Minister, 1968-70; elected Secretary General of the United Nations (1971 and 1976); first non-socialist president of Austria, 1986-92 – controversial because of his war-time past and in 1987 placed on the "watch list" by the US government. Currently president of the United Nations Association of Austria based in Vienna.

I am somewhat unhappy about talking of Jörg Haider as an Austrian phenomenon. You have these phenomenon in all of Europe. Here we have leftist parties and the right and parties of the centre. Why all this fuss? We are a democracy.

Q. Despite this some see his ideas for the Third Republic as outside the constitutional bow and pre-fascist and therefore not so democratic.

A. *Well if he commits a crime there are courts for this. I don't think we have more radicals than for instance in the United States.*

Q. But Haider's ideas for a so-called Third Republic, including a strong president and more direct democracy, for some go beyond the pale of the constitution and smack of a semi-authoritarian state.

A. *Well it's all a question of interpretation. You see sometimes I have the impression that we in Austria are screened in a way which is different from others. We have similar problems to other countries in western Europe. Just because of our past it's not correct to say we Austrians are rightist and dominated by the right.*

Q. Some of these criticisms are not from outsiders but Austrians themselves who say it will be a catastrophe if Haider gets into power. Are these people also misguided?

A. *I also went through this experience as such people spread lies to other countries. The calumny started here and was sent to the United States where it was picked up by journalists. Negative news is always more interesting than something positive. Wrong information was not checked and corrections not printed. We are all human beings even politicians and suffer from this calumny.*

Q. But are these lies? Haider has criticised representative democracy and some fear he wants a *Führer* state where parliament will count for little.

A. *The great majority of Austrians reject any kind of authoritarian rule. The disasters of the past have taught us this lesson. There are many other things more important than Haider. The situation is not so dramatic. We have the two main parties in coalition who are unable to decide on the budget and there has to be early elections. This is normal elsewhere in western Europe. There is always this discrimination against our little country. You should look at Belgium and Holland and of course Italy.*

Q. What do you think are the main changes necessary in Austria today?

A. *Social partnership which we have had for so long has produced an important and fruitful relationship. We have an excellent social security system but we can't pay for it. These things need reforming. This is all more important than Haider.*

Interview with Otto Habsburg
Vienna, 5 October 1995

Otto Habsburg, born 1912, eldest son of last Austro-Hungarian Emperor Karl. International president of the pan-European movement and since 1979 member of the European Parliament for Germany.

Q. What are your views on the "Third Republic" and Haider's proposals to change Austria?

A. *Well Austria has to be transformed in my opinion. It has to be transformed for the very simple reason that the institutions we now have in Austria are not suitable for the direction in which Austria is going. That applies to the European Union as well as to the internal development of Austria. You see all countries have to reform because although we have made tremendous progress since the nineteenth century in the economy (as a consequence of the scientific explosion) there have been few changes in political institutions and these no longer fit the present century. I mean this goes for most countries with the exception of the Fifth French Republic but this would be the only one.*

Q. The Austrian constitution is now 75 years old – what changes do you think are necessary? Would Haider's proposals for a "Third Republic" cause you concern?

A. *No, I think the most important thing would be to finally realise that we shall only have a modern state if we separate the thinking function from the administrative function. I have been in active politics now for many years and I can tell you the administrative function is today's search. You are snowed under and are left no time to think. In Austria we have also to separate these functions and get a system which is closer to the people than the one we have today. I am a great supporter of the direct election of members.*

Q. One feature of the Third Republic envisages more plebiscites....

A. *No, I am not for this. I am not for referenda. In Austria there was a referendum in 1978 on nuclear power – this was absurd, it was criminal because it showed the politicians were too cowardly to take a decision themselves. The people can't know. They have other worries. The politicians have been elected to work on behalf of the people. I'm for representative democracy but the election of* the representatives should no longer be submitted to the dictatorship of the parties but should be based on the direct election in the constituency. In the European parliament we still have different electoral systems, and I think the British one has many merits although I am not for their "first past the post" system and would prefer the French model. But you have to think of the grass roots. The party list system is wonderful for political science professors but that's all. It doesn't give you this contact with the people which is necessary.

Q. Would the idea of a strong president or head of state fit into your model for change?

A. *I would very much be in favour of that as I am a great supporter of the Fifth French Republic. If a president is elected by parliament he is just a third rate supporter of what the government wants. This is why when they asked me to be a candidate for the presidency in Hungary I said "no thank you." I don't want to be elected by the parliament. You have to be elected by the people. When a president is elected by the people there is the authority of the government and the authority of the president. You have to have a certain balance.*

Q. If Jörg Haider were to have a position in the government or held high office, do you think this would damage Austria's image abroad?

A. *At first yes, there is no doubt about that because he has been so much maligned and attacked so often that it would have an effect at first. But on the other hand if you think of his record as provincial governor of Carinthia, he's quite good in administration. So I think in the end he may do quite well.*

Q. But he lost the governorship there because of his remarks on the Nazis' employment policies.

A. *But that was a big balloon of nonsense to get rid of him.*

Q. Would you see him as a right-wing extremist?

A. *He is in his own element. A populist and we have a lot of attacks on populists because the established parliaments and parties don't want them. You have the same phenomena in Hungary and elsewhere. Some of these new figures in politics are simply a mirror which is presented to the people who are in power. They have become self-satisfied and very often corrupt.*

Q. Many on the right see Maastricht as a kind of super state. What would be your answer?

A. *Maastricht was on many psychological accounts a great error. It is too much written in bureaucratic language and people don't understand what it's all about. But on the other hand we have one thing which made me a defender of Maastricht. That is we have the principle of subsidiarity for the first time in an international treaty. This will be implemented in the courts. This is an important principle of decentralisation to place power at the roots. This is one thing we can get out of Maastricht.*
I am very much for decentralisation in Europe and for the regions. We shall have to have certain things in common: defence, security, foreign policy and finances but the all the rest should be done at the lowest possible level.

Q. Would Austrian neutrality be able to be maintained within this new Europe?

A. *I am against Austrian neutrality for the very simple reason it's absolute nonsense. I was not against it so long as Austria was on the border of Russia. But that has disappeared and now who do they want to be neutral against? Liechtenstein?*

Q. And NATO membership for Austria?

A. *Why not? I am very much for an extension of NATO as far as possible eastwards because we must not underestimate the dangers which come from Russia. We should extend Europe in general eastwards and we have little time to waste. This will be a great enrichment for the west. We ask too often what does it cost but not what does it not cost to extend Europe and have all these enormous potential resources of central and eastern Europe in the European Union.*

Q. Austria is in the European Union but has not held elections to the European Parliament yet....

A. *Because they're afraid of Haider. It's very simple.*
Austria is discrediting itself by delaying these elections. The Freedom Party in Europe is doing some excellent work.

INTERVIEW WITH JÖRG HAIDER
VIENNA, FEBRUARY, 1997

Q. You come from Bad Goisern, Upper Austria. How did you see Vienna as a child?

A. *Vienna really played no role at all. We were in the northern part of the Salzkammergut. The provincial capital of Linz was for us nearer. No one really thought of Vienna at all. Everyone was very concerned with their own region.*

Q. When did you first visit Vienna?

A. *It wasn't until I was 18 and of course I was very impressed with the splendid buildings and so on. It was another world.*

Q. You decided then to study in Vienna instead of Linz.

A. *No, instead of Salzburg. To begin with I wanted to study in Salzburg. I had rooms reserved and everything and then suddenly I made a spontaneous decision not to go. I wanted to study German and history in Salzburg and then I opted for law in Vienna. I thought well, you can do more with that. First of all I did military service for a year and then I went to Vienna in 1969 and in 1972 I started as an academic under Professor Winkler which really taught me a lot.*

Q. When did you decide to go into politics?

A. *It was in 1976 although before I had been leader of the Ring of Free Youth for Austria and then I gave that up since I was in disagreement with the policies of the federal party at that time. I withdrew from politics for a time until 1976 when my Carinthian friends took me by storm and said "we're beginning anew and we're going to reform the party and you should join us." I was attracted by this and said to Professor Winkler "what do you think?" and he said "it's a good opportunity. You can always come back to me; your academic career will still be open but it's good to get some practical experience and see how things really work. Give it a try – it's a challenge."*

Q. I believe you originally wanted to be an actor?

A. *As a school kid I really did dream of this. We had a stage and a school theatre in Bad Ischl which put on some very successful performances. We did Nestroy and Gerhard Hauptmann and Raimund. I had some lead roles in plays such as "einen Jux will er sich machen" and "Lumpazivagabundus."*

Q. Were your parents against an acting career?

A. *Well I wasn't really that serious about it – it was a schoolboy wish like wanting to be a train driver.*

Q. I think you may use some acting in politics?

A. *Of course the experience certainly wasn't wasted!*

Q. What was your school like – was it very strict?

A. *Firstly it was a private school which wasn't very common for Austria, but a private school where you could do public exams – but you had to pay 260 Schillings a month which was a lot then especially as my father didn't earn much and my mother wasn't employed. My sister was also studying. My parents even took out credit in order to finance our studies. It was a strict school, yes – but it didn't do any harm.*
The school was 9 kilometres away from where I lived, in the "Kaiser town" of Bad Ischl. It was really good – we used to go by train and very often in the winter when there was a lot of snow, it was late so we missed some classes. Generally though I liked studying especially Latin. Some subjects not so much like maths for example that annoyed me for a while.
The director of the school was a conservative ÖVP man, I think he was of Jewish descent and one of his closest colleagues was a pan-German nationalist. A funny sort of mixture.

Q. Your parents were former Nazis....

A. *Well, they were like hundreds of thousands of others, members of low rank with small functions in the party. My father was responsible in Upper Austria for organising the Reich apprentices' competitions for young people and my mother was in the League of German Maidens. After the war they were classified as small Nazis, whilst the really big ones sought their careers with the ÖVP and the SPÖ. They became the mayors and ministers and politicians for the province whilst the little people were penalised including my parents. They were banned from normal jobs and had to do work in hospitals or homes for the elderly cleaning up and doing menial tasks. My father was forced to work in the cemetery as a gravedigger. This didn't last very long though.*

Q. This was the American zone of occupation?

A. *Yes although I always had very positive impressions of the Americans. The first things we got were huge cans with delicious yellow cheese, milk and lots of other goodies.*

Q. And your grandparents?

A. *My granddad on my father's side was an innkeeper in Mondsee – I think he also had a butcher's shop. On my mother's side, the mother came from South Tyrol and the father was a gynaecologist who studied in London and worked there. He was a real Anglophile. He travelled a lot as a medical doctor, a real cosmopolitan or "globetrotter" as we would say today.*

Q. And as a student you were a member of the student duelling societies?

A. *Even in Bad Ischl. There were two – one was the catholic and the other was if you like "Freiheitlich" and this one I joined. The director's son who went to school with me went to the catholic society. We had a good friendship both personally and between our groups and competed with each other to see who could get the most members. It was fun.*

Q. Was this also a Protestant area?

A. *Partly, the "inner" Salzkammergut if you like was a retreat for Protestants during the Counter-Reformation. Especially in and around Bad Goisern there is a strong Protestant community and nearby, on a mountain you can reach in about two hours by foot, there's a so-called church, an enormous cave where the Protestants celebrated a service to commemorate this as a sanctuary. It's really interesting – you have to go through a dark passage for about 20 minutes in this cave right in the middle of a mountain and suddenly there opens up in front of you a huge vault just like a church and this is where the Protestants held their service.*
Salzkammergut is the heart of Alpine territory which many know from "Sound of Music" with its wonderful lakes and the famous salt works or the Hallstatt Lake at the foot of a glacier. One of the most beautiful mountains is the Dachstein which has inspired many songs like "Hoch vom Dachstein an, wo der Aar noch haust." ...and then there's the famous Erzherzog Johann Yodel. Erzherzog Johann was a native of these parts and for me is a symbolic figure and a model since he was a revolutionary against Vienna, against the Kaiser house although he was a son of this imperial family and was then in 1848 the first president of the Frankfurt national assembly which was concerned with the first constitution and an attempt to set up basic rights and freedoms. He was a real reformer. He then came back to Styria and fell in love with a well-to-do lady. He then put through many social reforms and was amazingly innovative and creative and also was a great mountain climber. My political model.

I can't think of so many contemporary models – I was once impressed by Helmut Schmidt, the former Social Democratic chancellor of Germany.

Q. And Kreisky?

A. *He was one of the last ones who really tried to reform this country and open it up. I talked with him just before he died. The account of the meeting is always disputed by the Social Democrats who are obviously annoyed that it took place at all especially as most of them never knew it happened. The then ambassador Gredler who knew Kreisky well and who was "Freiheitlich" requested a meeting. What sticks in my mind from what he said was that in the future there will no longer be a two party system but there will be a trend to a three party system, three roughly the same strength with the Social Democrats just in front but one must ensure, and this is what he expected of me, a dialogue between all is kept up because we in the FPÖ are an important element in Austria's democracy and therefore it was important to maintain discussions with all sides. And I think our struggle against political isolation is connected with this – we say it has nothing at all to do with Kreisky's legacy when the Social Democrats try to isolate us since that is exactly what he didn't want.*

Q. You have experienced many victories but also setbacks. Was there ever a time when you thought you would rather pack it in?

A. *There are always situations when you might say, "Do I really have to do this?" That was the case for example in 1986 at the party conference when there were some savage debates and after 7 hours I said to myself, "Do you really need to be insulted in this way?" But my friends encouraged me and said stick it out, it'll be all right. It was always like that and I have the advantage of having really good friends who have stuck with me through thick and thin and of course the solid backing of my family has been important else I would probably not have carried on with it.*

Q. Would you do it again if you had the chance?

A. *It's difficult to say since I've experienced it this way. But many personal chances in life are lost since there is an enormous concentration on specific goals. On the other hand I'm not sure if I would really be satisfied with a so-called normal life when you go home at four o'clock or whatever or as a university professor and you do your lectures and exams and that's it. It's not me. I need the challenges.*

Q. Was one of these "downs" after the 1995 election?

A. *Yes, for me personally it was since I really couldn't understand how dishonesty could be so rewarded in politics – that really disturbed me that this blatant lack of truth which was dished out to people in the socialist propaganda led to a great election victory but you see poetic justice comes a year later which wouldn't have come in this form if they hadn't lied to the people. There's always some good which comes out of bad.*

Q. In your speech at Krumpendorf you referred to the "Hafenstraßen" delegation in Hamburg – what happened there?

A. *I had been invited to a meeting of the Bund Freier Bürger with Mr. Brunner and a lot of people showed up at this square in Hamburg and the "Hafenstraße" that's where a lawless bunch have occupied a particular area of the docks and not even the police dare go in – it's a kind of "wild west." There are really radical elements, above all left extremists and they turned up and tried to start off fights and so on amongst the crowd. They then chucked everything possible at us on the speakers' platform stones, bricks and I don't know what – it was really nice. Anyway we carried on with the meeting and the police tried to give us some protection and took into custody some of them who had dangerous weapons – one just in front of me drew out a long knife. They were completely vicious characters. And then this guy in Krumpendorf who came from Hamburg apologised and said how embarrassing it had all been and really the people in Hamburg are quite friendly.*

Q. Then you spoke of someone who should have come the year before but couldn't land....

A. *That was Fasslabend, the Minister of Defence, who was down to be the speaker in 1994 but he got cold feet for political reasons and cried off but the official reason was that it was so foggy the plane wouldn't be able to land although in actual fact it was a lovely bright and sunny day so it wouldn't really have been a problem. Anyhow he showed up the year after.*

Q. What did you mean when you referred to the "decent" people – the former Waffen SS?

A. *I was talking about the older generation and then I mentioned the Freedomites who represent a kind of elite in this democracy since we will bring about change and without us this rigid system in Austria won't be broken. There were journalists from* Stern *and*

Profil there – no one got excited about the speech until months later they started to criminalise it.

Q. In a subsequent interview you referred to the memorial to the Soviet army still standing in Vienna....

A. What I was trying to point out was the contradiction we have in Austria when on the one hand people try to do away with war memorials dedicated to those who fell in the war from all political sides and religious denominations whilst on the other side we still have a Soviet war memorial in Vienna. This should be demolished in the post cold war era just as those people in countries once under communist oppression have got rid of Lenin and Stalin memorials.

Q. You have three large flags here in your office – one naturally enough the Austrian flag and then also the Star Spangled Banner....

A. And the third is the flag of the state of California which has a bear on it and reminds me of my home in "Bear Valley" Carinthia. I've always liked the Stars and Stripes. I'm a big fan of pretty flags and I think the US flag is really great.

Q. If you were in America would you be a Republican or Democrat?

A. More of a Republican I suppose but it's difficult to say. In economics I'd be more on the liberal side and for the market economy and for deregulation. In social and family policy etc., I'd be more on the conservative side.

Q. You studied at Harvard University last Summer....

A. Yes, privatisation policy. It was wonderful to be there and have some peace to think. It's a wonderful country – so uncomplicated in comparison with us and yet it's unconventional. Then I liked the way they study there and the staff student relationships are a real partnership and the professors usually come from the real world not like here where you often get professors who hold the same lectures for thirty years or more. It's what I imagined it should be like at university where professors have practical and up-to-date experience. There were experts who had been engaged in consultancy in privatisation in Eastern Europe and who had made reports for the World Economic Forum; real economic gurus. It's a different system from Austria of course but there students who get on and become famous put something back into their universities financially.

Q. Were you ever in Chicago? What was behind these posters "Vienna must not become Chicago?"

A. *I've been there but the placards had more to do with this historical image since the city used to be synonymous with organised crime and there are many politicians in Europe from different parties who use such slogans. But the city itself in the centre is a modern, fascinating city.*

Q. America is a multi-cultural society, something which you reject for Austria....

A. *I don't think you can stop it but you should not make an ideology out of "multi-culturalism." People understandably defend their own homeland and we should be able to do this in Europe and Austria. The Palestinians fight for their right to a homeland and the Kurds, so it's difficult to understand why in the heart of Europe through a wrong immigration policy the peoples are being substituted by others from outside. Also the speed and the volume of immigration is important and makes it difficult to absorb.*

Q. You visited the Wiesenthal Centre in America – where did you see your portrait?

A. *Right at the beginning of the exhibition which depicts the suffering of the Jews in the Holocaust. First of all there's a section on the violation of human rights in the whole world such as what the Americans did to the Indians and the execution of civil rights fighters and at the end of this there's a huge wall with lots of photos which serves as a warning that many politicians still plan to eliminate human rights in the world of today. There's Idi Amin and Saddam Hussein and then there's me. Then I asked the guide, "Who's that guy?" and he said "Oh, that's a dangerous right wing politician from Austria!"*
And then you go on further through a door and it's cool and dark and that shows the suffering of the Jews in the Holocaust. It's really well done but it's the overall impression – first of all outside with these photos and then you go straight into a section dealing with the extermination of peoples.
I also went to the Holocaust Museum in Washington but at least I'm not shown there. It's quite effective with a history of the family of Daniel. It's the story of little Daniel who with his family grows up in the ghetto in Eastern Europe and you feel his experiences as he goes through life. You see how he started in peace at school and see his little school bag in his room and so on and then they go to the ghetto and the entire family is split up. It makes quite an impression.

Q. Have you visited the exhibition in Austria on the German army?

A. *No, certainly not since I don't think I need to be told anything by former Soviet and East German communist historians. It's a falsification of history they've been carrying on as is clear since the Berlin Wall fell and we have access to the archives.*
I went to the Mauthausen concentration camp site once on a private visit but it's got nothing constructive to say in comparison with the Washington Holocaust Museum.

Q. If we switch to the future from the past, are you a follower of astrology?

A. *No, but I get all possible kinds of horoscopes sent. As an Aquarian I'm in good company with Reagan, Kreisky and Paul Newman. I think some things fit like the general characteristics. For example they love freedom and you can never pin down an Aquarian.*

Q. With Klima you have the chance of a co-operation with the Social Democrats.

A. *I see this just as normalisation, the beginnings of normal parliamentary practice. Klima can't suddenly throw Vranitzky's legacy out of the window. I think the SPÖ is working on a coalition with the Greens and Liberals. If that doesn't work they might consider something with us. When power is at stake the Socialists can be very flexible. You can feel this change in the provinces and the ÖVP will have to think about things since it has been really stupid the way it has looked on. It was always frightened to co-operate with us now they see the Socialists might do it. You have to understand we have had many negative experiences with the People's Party, as for example in Carinthia when it broke a pact with us and then supported a Socialist as provincial governor instead. Provincial governor of Carinthia would still interest me since I think the province is in a terrible state and needs important reforms.*

Q. How do you see things developing at the European level?

A. *I can't see how with over 19 million unemployed there will be a secure peace in the future in Europe. This has been completely ignored. Some countries may become successful but then they will say we're not prepared to finance the others.*
We are for a Europe of the nations where the nation state has still an important role to play since the nation states are the only corrective against the power of the multi-national concerns. The

nation states can develop to a large degree an autonomous economic policy for people in small and medium-sized firms.

Q. Is Austria a nation or a fatherland?

A. *Austria is a state nation and the fatherland is a broader concept consisting of different ethnic groups such as the classic minorities such as the Croats, Hungarians, Slovenes and Czechs and then the majority of Austrians who belong historically and linguistically to the German Volk and cultural community.*

Q. Your party is now very successful as a party of the workers....

A. *Actually a party of the working people, i.e., of those who are not employed in the protected sectors of the economy. Small tradesmen, farmers, workers, salaried employees – all are in our voting alliance. These people should no longer have to bear the brunt of cut-backs and with the opening up of these protected privileged sectors we have a big opportunity to help those who now vote for us. We could then have more chances for greater competition, more privatisation in electricity, radio and television, and the whole state sector.*

Q. How can you avoid the temptation of building up your own power network in place of the Reds and the Blacks?

A. *When I was governor in Carinthia I could see it will work if you want. You have to have jobs allocated on an objective basis. This was good for Carinthia and it wasn't important whether someone was a socialist or a member of the ÖVP or a party member at all, but the decision to appoint someone must be made on the basis of merit and qualifications. People want to be rid of this political pressure and they want an end to political abuse when it comes to getting accommodation or credit at the bank, and jobs etc. We want to see real competition between the public and private sectors instead of a monopoly of politicised housing cooperatives which hand out apartments. This whole party book system must be a thing of the past. If someone wants to join us or have an Info card because they want to know more about our aims or support us – OK. But if they think it's a ticket for their personal enrichment they're wrong.*

Q. What is the "Haider phenomenon"?

A. *A lot of things have to do with destiny. I wanted to reform this country since it really annoyed me how it was politically divided up. Just at the time I took over the party leadership I became financially independent. I always operated without any network or old school tie connections. But I got this financial independence from my uncle which was completely unexpected since I had not had that much contact with him. I think he sensed that I was taking over an important job and said to himself "he needs some help." He was a big fan of my politics and must have thought "he needs support, he's got a family, children to bring up and this will ensure he won't be open to corruption." You need this independence in politics to follow goals with determination otherwise you can easily be bought off. There were times I experienced when powerful socialist politicians said to me, "stop following these policies against us else we'll see to it that you won't get on in Austria." They felt themselves to be complete rulers.*

Q. But many things have changed and it's not quite so "feudal."

A. *Because of us – without us this wouldn't have taken place. The* Proporz *is on the way out, the absolute majorities are gone, that's very important so all these little functionaries are no longer so powerful. The Reds and Blacks can no longer do what they want. Then we reduced their privileges and so step by step we are getting there. We've got a long way to go before we can say Austria is a normal democracy.*

Q: If you achieve that then you'll be superfluous.

A. *I shall have been a long time in politics by then.*

Q. Will you sort your wine cellar out like Vranitzky then?

A. *I certainly won't be in the wine cellar like Vranitzky but there are other hobbies for me like more sport, mountain climbing. I don't have enough time for this. Last weekend I went on 60 kilometres cross country skiing. Marathon running I also enjoy since I think a lot and often in my mind write my speeches for parliament when training. Also I'd like to write another book sometime – I'm not sure what about.*

INTERVIEW WITH SIMON WIESENTHAL
VIENNA, MARCH, 1997

Born in Galicia in 1908, Wiesenthal studied architecture at the universities of Prague and Lemberg. In all he suffered in twelve concentration camps until finally liberated in May 1945 in Mauthausen, Upper Austria by the Americans. For two years he worked for the US War Crimes Office. In 1947 he founded a Documentation Centre in Linz to collect material on the fate of the Jews and their persecutors. After the capture and trial of Adolf Eichmann, Wiesenthal opened a new centre in Vienna which he has directed since 1961. Over 1,000 Nazis were tracked down and brought to trial through the research of Wiesenthal.

Q. Do you think Haider is dangerous, especially if he came to power?

A. *It's not easy to say and it depends on the situation. At the moment around 70-75 percent of the people are against him and this will not change much. But he has the chance to get into power.*

Q. What would he do if this happened?

A. *Look, he is not a neo-Nazi; he's a radical – he could be a radical left or radical right because he wants to change our society. When he was first in parliament many years ago he was more left than the Social Democrats. His whole politics has one aim and this is for the people in pubs for whom he carries on his political propaganda. The problem is that the coalition of the ÖVP and the SPÖ has lost contact with the people.*
He gets a lot of votes from the SPÖ and from the workers. It all started in 1986 with the election of Waldheim as president. The Socialists were against him. Contacts were made with the World Jewish Congress to get help against Waldheim but he won. Soon after Haider became leader of the FPÖ and kicked out Steger. The ÖVP thought they would keep the Waldheim voters but they went to Haider at the next election. Haider knew these votes for Waldheim were not for him as such, but were votes against foreigners and against the World Jewish Congress. So today Haider can say there are 300,000 people out of work and 300,000 foreigners in Austria. It's false logic but it stays in the minds of people. Now the coalition runs after Haider.

Q. Haider's portrait hangs in the Wiesenthal museum in Los Angeles apparently along with Idi Amin and Saddam Hussein. Would you compare him to these people?

A. *No, that's far too simple. I've talked to them over there about this. They promised me in the museum that they would change it but I've not been there for two years. OK, you could compare him with Le Pen or Schönhuber but not with these people who eat human liver.*

Q. How do you think Austria has come to terms with its past?

A. *Austria claimed after the war it was the first victim of Hitler. It was – the country that is but not the population. Austrians were in the Nazi party (in leading positions) and in the SS and were over-represented in the Holocaust. They were very proud that Hitler was an Austrian.*
    In 1938 in Carinthia many Socialists went over to the Nazi party and after the war they went back again. They were always against foreigners there and they don't like people from the other side of the border, i.e., from Slovenia. During the war many crimes were committed by Austrians in German uniforms in the Balkans.
    Haider has his big estate now in Carinthia which was Jewish property. The owners were Italian Jews who bought it in the 1930s. The property was "ex-Judaised." Haider later got this as a donation from a relative and its worth around 100-150 million Schillings. He's one of the richest men in Austria.

Q. What of Haider's parents...?

A. *They were Nazis although not very big ones, but Robert Haider was in the party at a time when it was forbidden in Austria. Haider was educated by them and many Nazi slogans he utters he remembers from home. Things like punishment camps – for the Nazis everyone in concentration camps was a criminal.*

Q. For the European elections a Jew, Peter Sichrovsky, stood for the FPÖ list. Do you think this means the party is open...?

A. *We have no monopoly on Jews. But I have to say one thing – Haider never said anything against Israel and has never said anything anti-Semitic. Never.*

Q. Sichrovsky was invited to attend a Jewish cultural week in London because of his writings, but the invitation was cancelled when he became politically active....

A. *I know. Look. About twenty years ago I was invited to attend a meeting of the academic association "Teutonia" in Germany. They're not extreme right but right. I was surprised but I agreed. Everyone told me not to go because they are against me. But I said there's no point talking all the time about anti-Semitism only to*

Jews. So I went and gave a talk. There was some applause (not much but some) and then discussion. The first contributor said, "Mr. Wiesenthal, we know you. You eat a Nazi for breakfast and a Nazi for lunch and a Nazi for supper." I interrupted him politely and said, "This is not true, you see I don't eat meat from pigs!" After a short silence there was loud applause and we had a good discussion. I still have contacts with people from that meeting.

Q. Last year there was a meeting in Vienna of right student duelling fraternities.

A. *Yes, duelling is their hobby. This has been going on for generations.*

Q. Do you think they could be dangerous?

A. *I think not. For the leftists yes since everyone not left is a danger. I don't like anti-Semites but I don't like philo-Semites. In the interwar period there were Jewish duelling fraternities.*

Q. There's a controversial exhibition on the Wehrmacht....

A. *It's a compete generalisation. You can't accuse all those who were in the Wehrmacht of these crimes. We Jews were victims of generalisations for centuries. We should not fight back with the same.*

Q. How do you think Kreisky would have dealt with Haider?

A. *Kreisky was only interested in getting a majority. He got a lot of votes of former Nazis. Why? Because he said he was not a Jew and the religion of his parents was of no concern to anyone else. He opened the way for a coalition between the Socialists and the FPÖ. He had former Nazis in his cabinet and I published their party numbers. This was the start of our dispute. In all parties there is some verbal anti-Semitism. Socialists and Communists and all of them.*

APPENDIX 1

TABLE A1
CHANCELLORS OF THE SECOND REPUBLIC

Year	Name	Party
1945-53	Leopold Figl	ÖVP
1953-61	Julius Raab	ÖVP
1961-64	Alfons Gorbach	ÖVP
1964-70	Josef Klaus	ÖVP
1970-83	Bruno Kreisky	SPÖ
1983-86	Fred Sinowatz	SPÖ
1986-97	Franz Vranitzky	SPÖ
1997-	Victor Klima	SPÖ

FIGURE A1: STAGES OF LEGISLATION

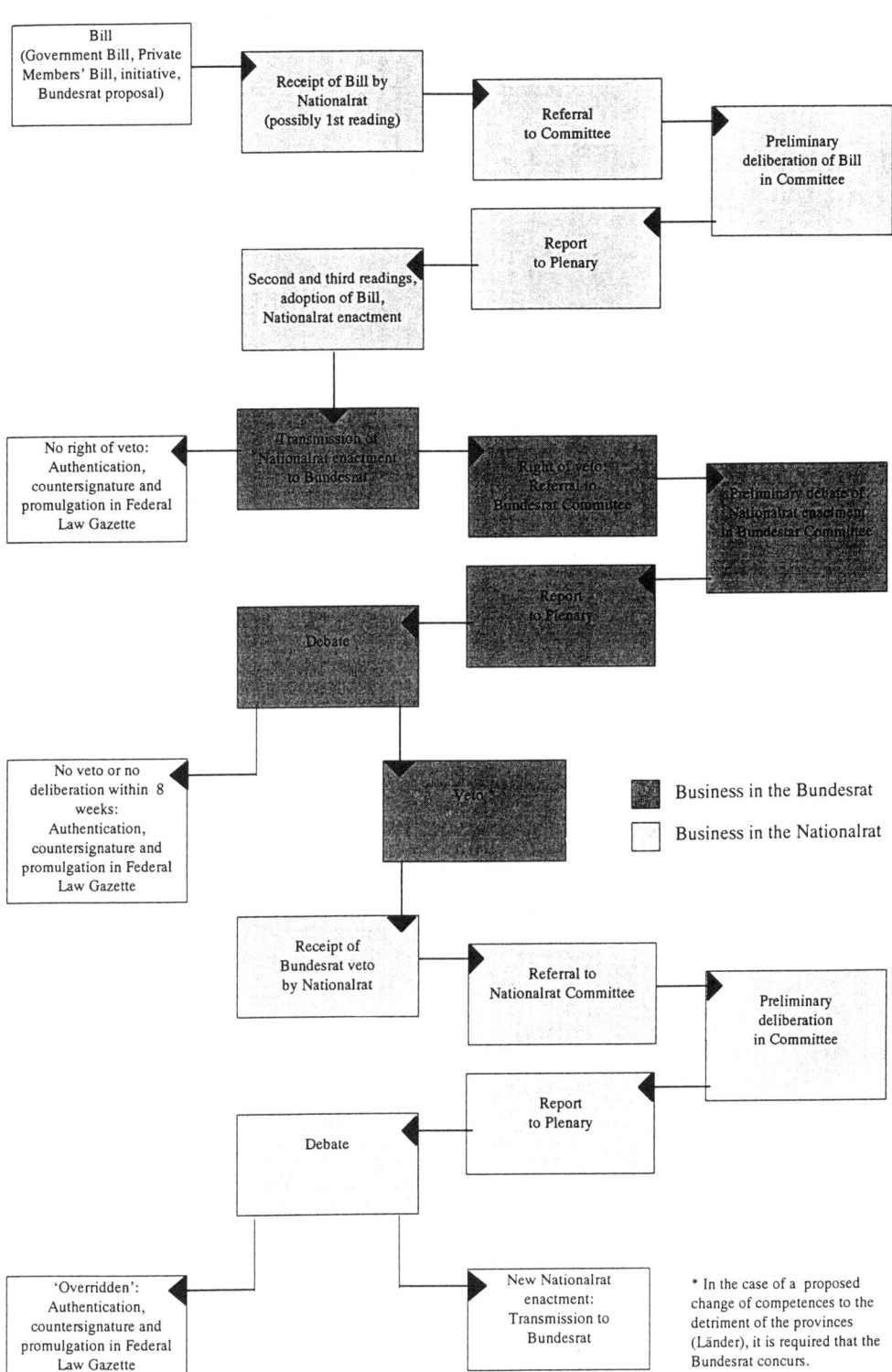

Source: *The Austrian Parliament*, Vienna, 1992.

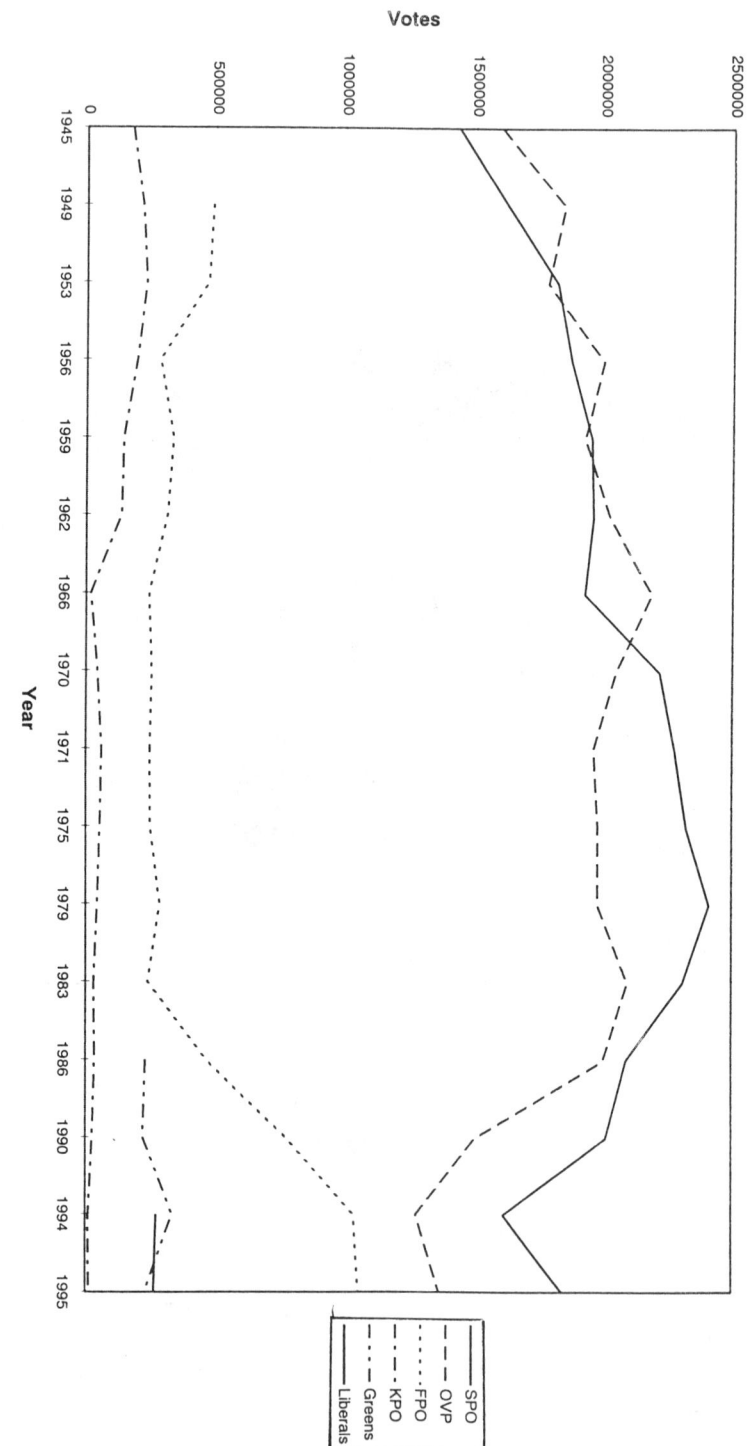

FIGURE A2
ELECTORAL PROFILE OF THE PARTIES, 1945–1995

FIGURE A3

ELECTION RESULTS FOR THE MAIN PARTIES (NATIONAL COUNCIL ELECTIONS EXCEPT FOR EUROPEAN PARLIAMENT ELECTION, EU 96)

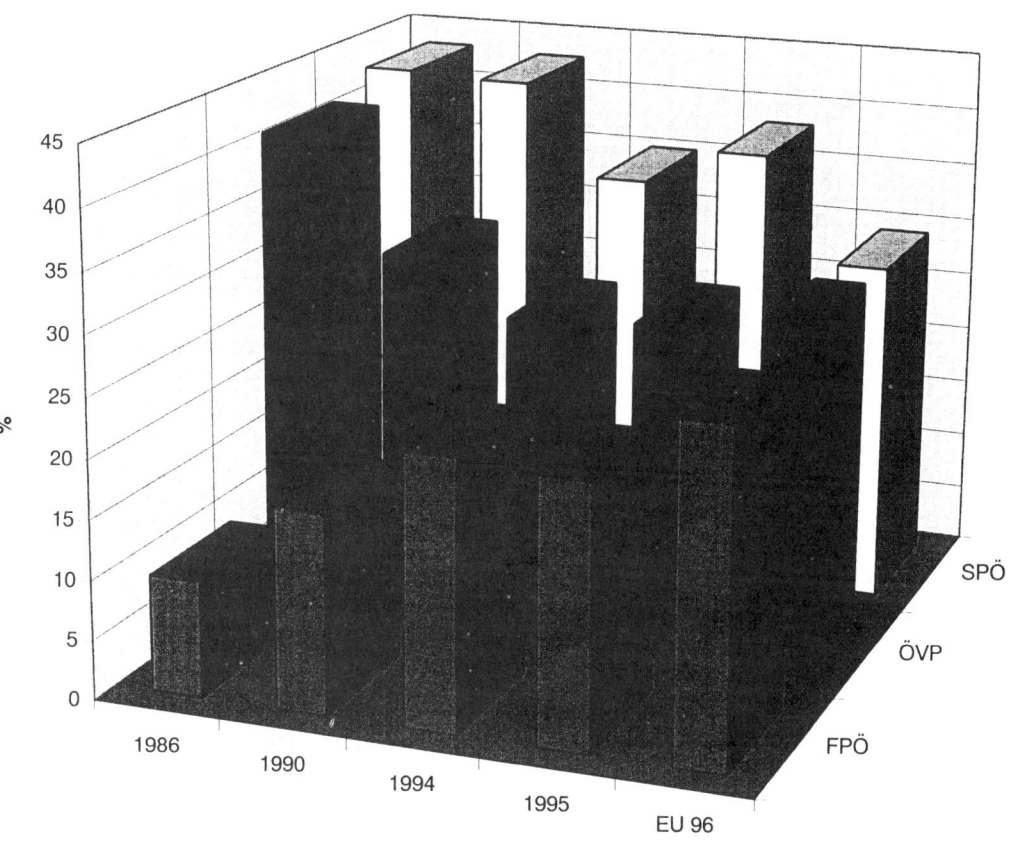

APPENDIX 2

BRIEF PROFILES OF OTHER MAIN PARTIES

THE SOCIAL DEMOCRATIC PARTY (SPÖ)

The Socialist Party of Austria (SPÖ) was newly founded in 1945. It adopted the more respectable name of Social Democratic Party after the collapse of Communism but this was one of the few reforms of any note.[1]

The party was the junior partner in the Great Coalition with the People's Party (ÖVP) 1945-66. It then had a brief spell in opposition during which Bruno Kreisky implemented a reform drive to prepare the party for government.

Since 1970 the SPÖ has emerged from every general election as the largest party in the lower house of parliament. It has held the chancellorship since then. Under Bruno Kreisky as chancellor (1970-83), Austria followed a policy of full employment and expansion of the welfare state. From 1983-86 the SPÖ entered a coalition with the FPÖ under the liberal Norbert Steger as vice-chancellor.

Franz Vranitzky became chancellor in 1986 and took over as party leader in May 1988. Initially the Vranitzky factor was able to stave off new challenges to the party's predominant position in Austrian politics posed by Haider's FPÖ and the Greens. He stood for a pragmatic kind of Social Democracy which accepted some degree of privatisation whilst preserving the role of the state in areas such as education and health. His main *raison d'être* was to stop Haider. This proved insufficient as an exciting agenda to attract new or floating voters. After the shock result in the 1994 election he looked weary and seemed unable to provide much inspiration for a demoralised party. He made a surprise come-back with the 1995 election which was called prematurely but this was insufficient to quell internal rumblings that the party had lost its way. The crushing result of the election to the European parliament (October 1996) was a factor in his resignation in January 1997. His successor was his finance minister, Viktor Klima.

Membership of the party has been on the decline (see Figures A5 and A6). Once under Kreisky the party could enroll around 10 percent of the population as members. Today it has under 500,000 or just over 6 percent of the population. The days of the mass membership party are over but even so many in the SPÖ find it hard to adapt to the new age politics. Just as worrying for the party is the lopsided age structure with most members coming into the category of pensioners, elderly or middle-aged. One in three members are over 60 and only around 50,000 are under the age of thirty. Young people simply do not find it attractive or necessary to join the party. The party's basic programme dates from 1978. It should have been updated by now, but lethargy has always won through.

THE AUSTRIAN PEOPLE'S PARTY (ÖVP)

The Austrian People's Party was founded in April 1945 as a successor to the pre-war Christian Socials. Both the ÖVP and the SPÖ pursued the politics of reconciliation in the Second Republic. The horrors of civil war between the two camps in 1934 had weakened the country. The Nazi *Anschluß* followed by war and four power occupation brought former political opponents together to work for the good of Austria. The Socialists followed a reformist course away from old Marxist dogma while the ÖVP, for its part, broke with the political clericism of the First Republic. The party stressed its independence of the Catholic Church as a Christian party. Unlike the First Republic these two camps, the Reds (SPÖ) and Blacks (ÖVP) worked closely together in coalition government.[2]

The FPÖ attacks both these parties as the "old" parties of Austria. This refers not simply to age or historical origins but also implies that these two political groups are a spent force whose heyday is over. For the FPÖ the Great Coalition spawned an extensive "old boy network" which divided power neatly between the two sides not only in politics but also administration, banks and nationalised industries. The *Proporz*, proportional allocation of posts in accordance with electoral strength, paved the way for the abuse of power and corruption, according to the Freedom Party which was left out in the cold during this time.

The People's Party was the strongest party until 1970 as the "chancellor party." Since then it has been on the defensive constantly changing leaders and strategy in an effort to win back some ground. One of its main problems has been the difficulty in co-ordinating the different organisations in the party: the Federation of Workers' and Employees (ÖAAB); the Farmers' Federation (ÖBB); the Economic Federation (ÖWB), in addition to the Women's Movement (ÖFB), the Young People's Party (JVP) and the Senior Citizens' Federation (ÖSB). These all have diverse aims and priorities and are also organised in the nine federal provinces. The leaders of the sub-groups and the provincial parties can sometimes be political heavyweights. In the past there have often been conflicting statements issued by these groups and internal battles have been fought publicly giving the impression of a divided and unsure party. There has been constant talk of dissolving these sub-groups altogether but nothing concrete has transpired.

In the spring of 1995 after yet another election debacle, Wolfgang Schüssel, aged 50, took over as leader from Erhard Busek. Schüssel who also became foreign minister and vice chancellor in the coalition government was regarded at first as more co-operative by the Socialists. He tried to appeal to young voters and women and set out to attract back voters in the towns who no longer saw the ÖVP as sufficiently modern. Schüssel himself projected the image of an urbane but athletic liberal intellectual.

The party's programmatic position derives from Christian Democracy but is fuzzy at the edges. Schüssel, like his predecessor, talked of the ÖVP as the centre force. It was hoped this would win the "common sense" voter who found Haider or the Greens too "extreme." The ÖVP supports a traditional view of the family and Christian values. It is attacked by the Liberals for being too "conservative" or reactionary on issues such as homosexuals, the position of women in society and law and order. For the Greens too, it has become a typical "law and order party" and follows a hard line on immigration. Its liberalism is evident in economic theory and the ÖVP supports a balanced budget, less bureaucracy, privatisation and deregulation. The 1995 election proved to be a gamble which went wrong and only modest gains were made. After the 1996 elections to the European parliament the ÖVP emerged as the strongest party but soon found itself again on the defensive.

THE LIBERAL FORUM

The Liberal Forum is the youngest of the parties currently represented in parliament. It was formed in February 1993 after five parliamentary delegates seceded from the FPÖ. Heide Schmidt, former general secretary of the FPÖ and the party's presidential candidate in 1992, became the chairperson of the newly formed Liberal Forum. The party is often accused of being a "one woman show" and the personality of Schmidt has been an important factor in mobilising support. Many voters, especially women, were impressed by her courage for eventually breaking with Haider and going it alone.[3] The party drew up a platform which shocked many conservatives with its liberal views on abortion, drugs and homosexuality (see Figure A4 for a relative public perception of the position of the parties). A prominent industrialist, Georg Mautner Markhof, broke with Haider and joined Schmidt but in 1995 left the Liberals because of this concern with so-called peripheral policies and an overemphasis on women's issues. In May 1996 he went back to the FPÖ. The Liberal Forum also had to respond to charges that it was "anticlerical" after it proposed to review the role of religious education in schools. It called for a renegotiation of the Concordat with Rome signed in 1933. It also proposed that the crucifix in schools and the court room should not be mandatory. In 1995 Schmidt announced that she had left the Catholic Church.

The 1994 election was a crucial test for the new party and Schmidt could feel well pleased with the results which saw the group returned to the national council. She came across, along with Green leader Ms. Petrovic, as the "power woman" championing the interests of women.

In parliament Schmidt's group operated a policy of loyal opposition. Whilst the Greens and the FPÖ opposed the government's pro-Europe course, the Liberals were in favour and supported the coalition in a call for a "yes" vote in the 1994 referendum. The Liberal Forum could claim it put the country before party but it was ridiculed as the "imperial opposition." It ran the danger of looking like an appendage to the coalition and lost momentum and profile. Some voices were raised within the party against Schmidt's alleged authoritarian style of leadership. Arguments were often too abstruse and intellectual and the provinces in particular were resentful of the Vienna "coffee house" liberalism. Schmidt was especially fond of podium discussions and press conferences in high class coffee houses in the capital. The provinces felt neglected but early elections in 1995 gave the party other concerns.

Before the 1995 election, Schmidt tried to regain some of the old punch of the young party which had inspired many young people. She defined the Liberal Forum as the party of the "offensive centre." She rejected a right of centre bourgeois bloc (ÖVP + FPÖ), and any alliance with the FPÖ, a party she regarded by now as hostile to human rights. The Liberals found the law and order platform of the ÖVP unacceptable. In an effort to combat terrorism, the ÖVP wanted to put through legislation involving greater powers for the police. Electronic eavesdropping, wider covert surveillance and a computer database on all Austrians were proposed. For Schmidt this undermined the foundations of the liberal state and represented a threat to civil liberties. The Liberals oppose state controls, a regulated economy and an oversized bureaucracy. The 1995 election promoted the party to fourth place above the Greens but the overall performance was disappointing. Internal splits and defections handicapped the party before the European parliamentary elections in 1996. The loss of drive and dynamism is currently a problem for the party.

THE GREENS

The Greens have been represented in parliament since 1986. Their members were activists in the successful campaign to prevent Austria's first nuclear power station from going on-stream in 1978. Subsequently they could be found demonstrating against the destruction of one of central Europe's last riverside meadow lands near Hainburg in Lower Austria. They, along with other activists including students, chained themselves to trees and threw themselves in front of the bulldozers. Violent and ugly scenes followed in the subsequent clashes between the protesters and the police. The trade unionists were concerned with jobs and argued for the project. A cooling off period saw a growth of public sympathy for the protesters and the government politicians finally backed down.

The early Green movement was split into a leftist group, and a bourgeois wing which laid less stress on social issues and feminism. Sectarian disputes blighted the electoral chances of the infant Green parties. Parliamentary work presented a different challenge for these grass root activists, suspicious of established parties and political leaders.

The party put up candidates for the presidential election in 1986 with Freda Meissner-Blau and in 1992 with professor Robert Jungk, a futurologist.

The reluctance to agree on a particular spokesperson was regarded as a handicap by political analysts. The birth of the Liberals under the leadership of a woman, Heide Schmidt, posed a challenge for the Greens who were stagnating. During the 1994 election campaign a new image was styled under Madeleine Petrovic who impressed voters with her well prepared television appearances but many Greens were uneasy that substance was being sacrificed to glamourous media presentations.

The Greens criticise "quasi-authoritarian, nebulous decision-making structures fostered by the existence of oligarchies and undeclared centres of governmental influence and seek to restore openness and public scrutability to government." The Greens want more opportunities for women in society and for those less well-off with families. They campaigned for a "no" vote in the 1994 referendum on joining the European Union, fearing a decline in environmental standards and an erosion of Austrian neutrality. The party is opposed to joining NATO and abandoning neutrality.

Under Petrovic the Greens built up a professional and respectable organisation. They are still vulnerable to charges from the ÖVP and FPÖ that they are too close to left extremist activists. For some veteran Greens like Ms. Meissner Blau, however, they have become a normal, even too normal, established party.

One of the most effective anti-Haider agitators is Peter Pilz who as a teenager was impressed by the guerrilla leader Che Guevara and communist Cuba. After a long stint with the Socialists he switched to the Greens and made a name in the Vienna City Council. He has the reputation of being a polariser and populist who can verbally slay a political opponent with merciless relish. Another Green leader, Johannes Voggenhuber, warned the electorate before the 1995 election against voting for Haider since this would pave the way for "neo-fascism."

After the disappointing election of 1995, Petrovic disappeared from the limelight and was replaced by Christoph Chorherr. His father had been the editor of the conservative daily *Die Presse* and, for some orthodox Greens, Chorherr junior was suspect because of his bourgeois background. His enthusiasm for the Internet and in-line skating cast him in the role of a distinctly modern politician with energy, drive, optimism and idealism. Articulate and educated, he could talk excitedly of his goal to change Austria but it was doubtful if his message came across to the older generation worried about crime and immigration. Since the 1995 election the Greens have looked just as divided and disorientated as in their infantile days over a decade ago.[4]

NOTES

1. See M. Sully, *Continuity and Change in Austrian Socialism* (New York: Columbia), 1982.

2. M. Rauchensteiner, *Die Zwei. Die Große Koalition in Österreich 1945-1966* (Vienna: ÖBV, 1987).

3. For a discussion of Liberalism and Nationalism in Austria, see G. Sperl, *Liberalismus gegen Nationalismus* (Vienna: Passagen, 1993).

4. For a discussion of the programmatic development of the Greens in the last ten years see R. Christian, "Die Entwicklung der Grünen," in Khol-Ofner-Stirnemann, eds., *Österreichisches Jahrbuch für Politik '95* (Vienna: Verlag für Geschichte und Politik, 1996), pp. 243-260.

APPENDIX 2 235

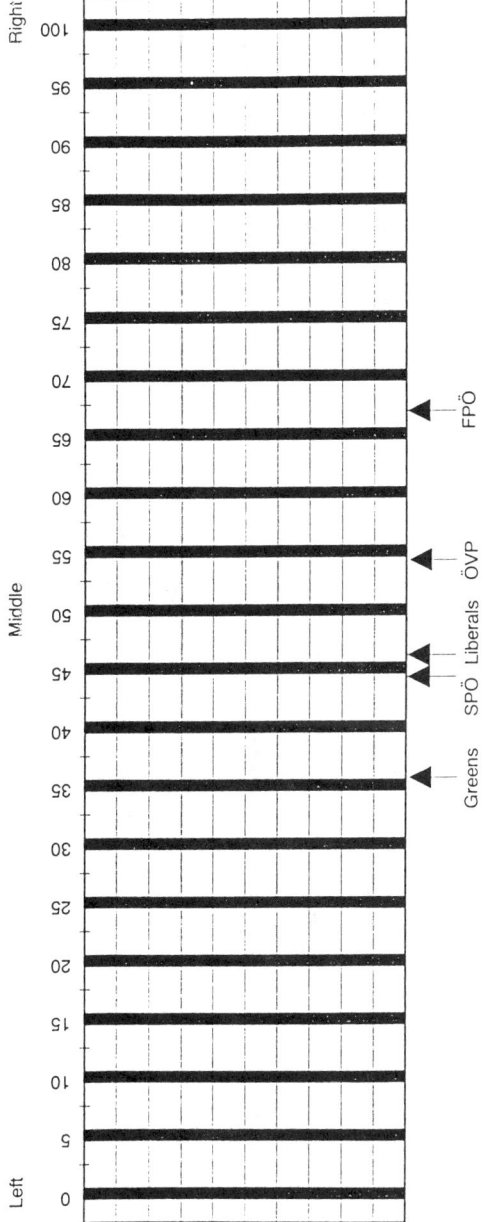

FIGURE A4
PUBLIC PERCEPTION OF POSITION OF PARTIES

Source: Kirschhofer-Bozenhardt, A. "Das Parteiensystem auf der Meinungswaage" in *Freiheit und Verantwortung*, Freiheitliche Akademie, Vienna, 1996.

FIGURE A5
SPÖ MEMBERSHIP 1945-1995

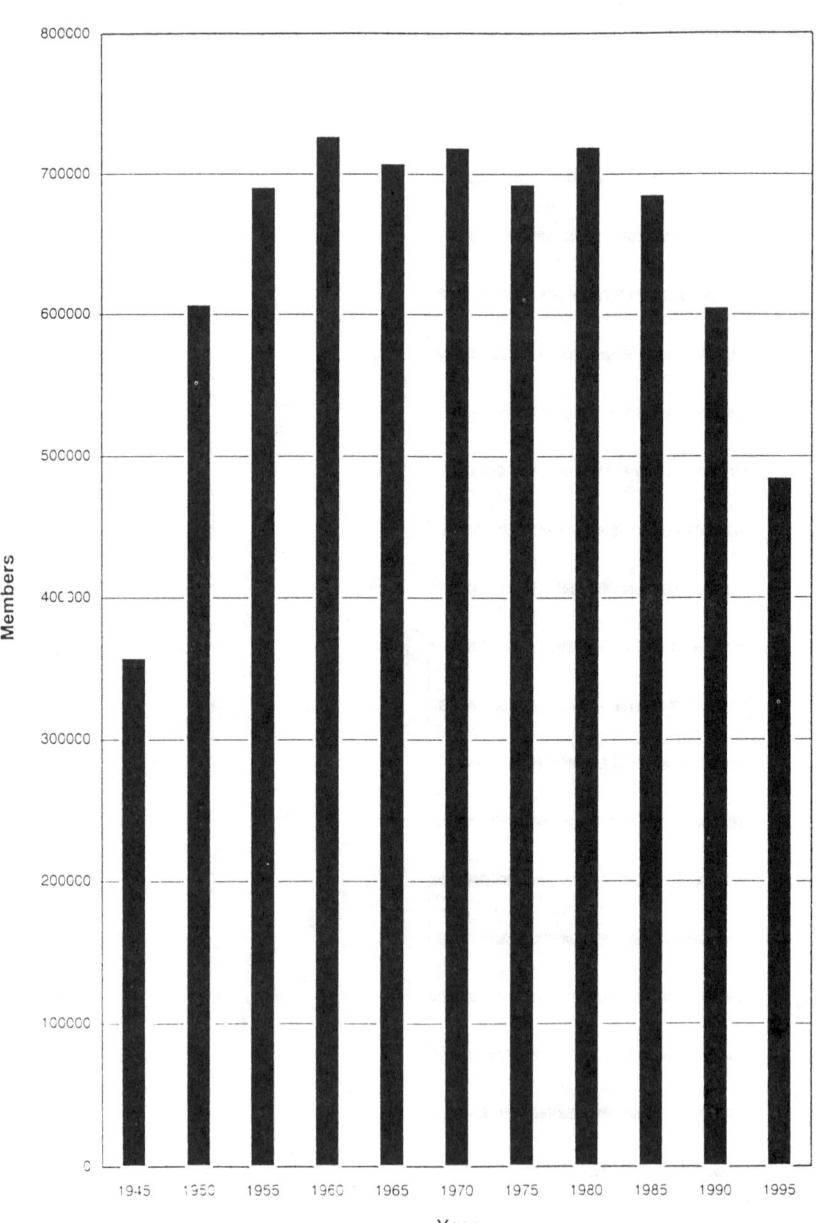

Source: *SPÖ Yearbook '95*.

APPENDIX 2 237

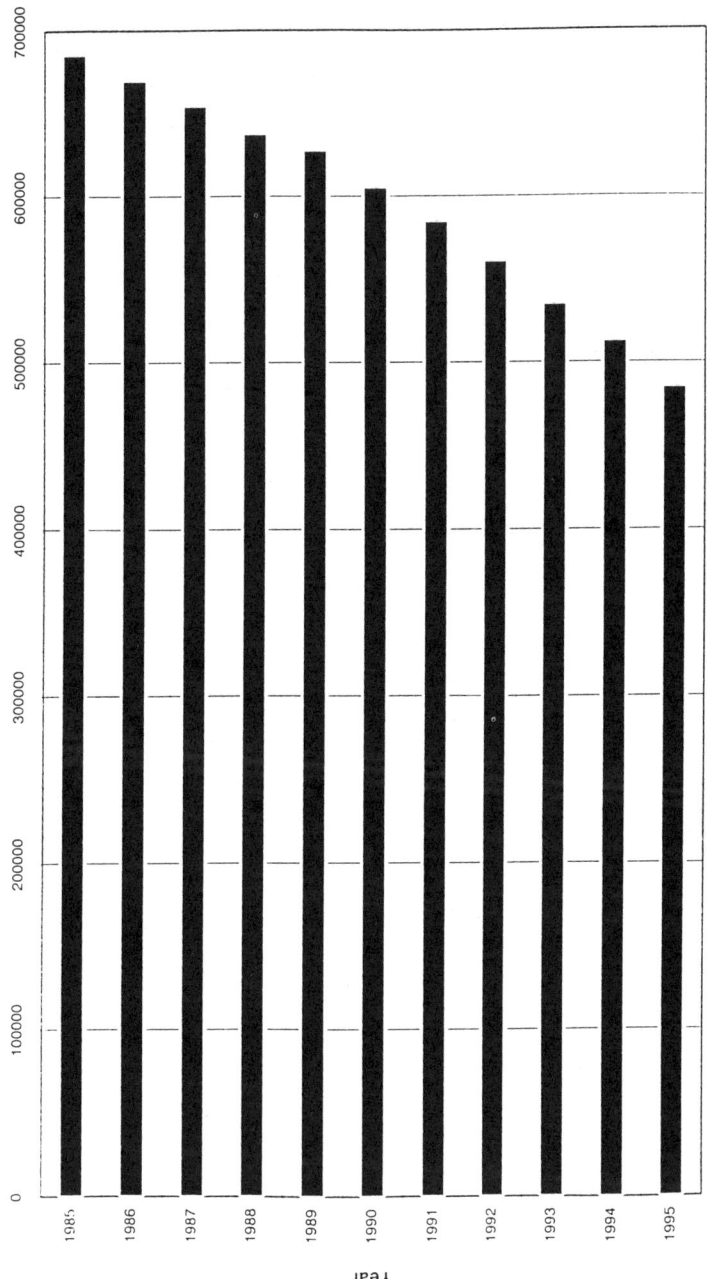

FIGURE A6
SPÖ MEMBERSHIP 1985-1995

Source: *SPÖ Yearbook '95*.

ABOUT THE AUTHOR

Melanie A. Sully was born in Bristol, England in 1949. She studied politics at the universities of Nottingham and Leicester and received a Ph.D. from the University of Keele. Dr. Sully was a lecturer at Staffordshire University and has been a regular contributor with articles on Austria for *The World Today*, Chatham House, London. She was a guest professor in political science at Innsbruck University (1988-91) and then in contemporary history at the University of Vienna. Since 1992 she has lectured at the Diplomatic Academy in Vienna.

Her previous books include: *Political Parties and Elections in Austria, Continuity and Change in Austrian Socialism* and *A Contemporary History of Austria.*